GLOUCESTERSHIRE AVIATION

A HISTORY

GLOUCESTERSHIRE AVIATION

A HISTORY

KEN WIXEY

ALAN SUTTON PUBLISHING LIMITED

First published in the United Kingdom in 1995 by
Alan Sutton Publishing Ltd · Phoenix Mill · Far Thrupp · Stroud · Gloucestershire

Copyright © Ken Wixey, 1995

All rights reserved. No part of this publication may be reproduced, stored in a retrieval system, or transmitted, in any form, or by any means, electronic, mechanical, photocopying, recording or otherwise, without the prior permission of the publishers and copyright holder.

British Library Cataloguing in Publication Data

A catalogue record for this book is available from the British Library.

ISBN 0–7509–0747–9

Library of Congress Cataloging in Publication Data applied for

Typeset in 10/12pt Times.
Typesetting and origination by
Alan Sutton Publishing Limited.
Printed in Great Britain by
Butler & Tanner, Frome, Somerset.

Contents

Introduction — vii

Chapter 1.   HISTORICAL SURVEY — 1
- Three Pioneers — 1
- Success Stories — 4
- Airfields — 8

Chapter 2.   GLOSTER AIRCRAFT — 16
- Mars 1 (Bamel) — 18
- The Racing Seaplanes — 21
- Gannet — 25
- Grouse — 26
- Grebe — 27
- Gamecock — 29
- Gamecock Derivatives — 31
- DH9As and Goral — 33
- Goring — 34
- The Big Move and Prototypes — 36
- Gnatsnapper — 38
- Gloster AS31 Survey — 39
- TC33 – Gloster's Jumbo — 41
- SS18/19 and Gauntlet — 42
- Gloster F5/36 (TSR 38) — 45
- Gloster F5/34 Unnamed Fighter — 47
- Gladiator — 48
- Gloster's Hawker Legacy — 51
- Gloster F9/37 — 57
- E28/39 and E1/44 – Gloster's Single-engined Jets — 58
- Albemarle — 60
- Meteor — 62
- Javelin — 66

Chapter 3.   BRISTOL AEROPLANE COMPANY — 71
- Bristol Boxkite and Early Birds — 76
- Scout — 79
- Bristol Fighter — 81

	A Bristol Miscellany (One)	82
	A Bristol Miscellany (Two)	87
	Bulldog	92
	By Jupiter	93
	Bombay	95
	Blenheim	96
	Beaufort/Beaufighter	98
	Buckingham/Buckmaster/Brigand	100
	Brabazon	101
	Freighter/Wayfarer	104
	Sycamore	105
	Britannia	106
	Type 173/Belvedere	108
	Type 188	109
	Type 221	110
	Concorde	111
	Bloodhound (SAM)	113
	Contracts	114
Chapter 4.	PARNALL'S OF YATE	115
	Panther/Puffin/Plover	119
	Possum	121
	Pixie	122
	Perch and Pike	123
	C10/C11 Autogiros	124
	Peto	125
	Imp and Elf	127
	Pipit	128
	Prawn	129
	Parasol	130
	G4/31	131
	The Heck Series	132
	Contracts	133
Appendix 1.	TECHNICAL DATA	
	Principal Gloster, Bristol, Parnall Aircraft	136
Appendix 2.	GLOSSARY	142
Bibliography		144
Index		146

Introduction

Among the many achievements credited to man this century, one of the greatest is his conquest of the air. In this field the county of Gloucestershire has played a most important part as various companies within its boundaries kept pace with the latest developments in aviation.

It took men centuries to emulate the birds, with strapped on wings, hot-air balloons, airships and gliders. Then the Wright Brothers fitted a small 4-cylinder engine to their biplane, and made man's first powered flight in an aeroplane on 17 December 1903 at Kitty Hawk, North Carolina.

Initially the British public paid little heed to the early aviators, but after Frenchman Louis Bleriot crossed the English Channel in his monoplane, UK public interest in aeronautical developments increased rapidly. With the outbreak of the First World War in August 1914, aeroplanes were inevitably developed as weapons of war. Engine power, reliability and sophistication increased as guns and bombs were added, and many airmen became public heroes. In the early post-war years some of these would prove that flight to anywhere in the world was possible, even a non-stop crossing of the Atlantic.

In Gloucestershire, Bristol Aeroplane Co., Gloster Aircraft Co., Parnall and Sons Ltd and George Parnall & Co. became well known and respected names in British aviation circles. However, the county also had its own individual pioneers whose efforts helped substantiate aeronautical achievement in the area. Some received little notoriety, others became well known, but they all revealed the spirit of enterprise prevalent among men of the county in the early and developing days of flying.

This book would not have materialized without the ready help of the following organizations, companies and individuals to whom the author is extremely grateful: RAF Museum; FAA Museum; MoD (AHB); Peter Hicks (Rolls-Royce Heritage Trust); Peter Pavey (Rolls-Royce AITD); Rolls-Royce plc, Filton; Westland Helicopters Ltd; Sir George White; Shell UK (Aviation Div.); Royal Aeronautical Society; British Aerospace (Filton); T.I. Jackson Ltd; *Gloucester Citizen*; *Gloucestershire Echo*; Russell Adams FRPS; John A. Long ARPS; David Charlton (BAe Airbus Ltd); Richard Riding; Harry Lane (YDOHP); Mike Goodall (Brooklands Trust); Peter R. March; Brian Pickering ('MAP'); Philip Jarrett; Mr & Mrs A.R. Hacker; Mrs M.A. Hedges; Bob Wilcock; Cyril Palmer; Clive Dodge; Mr G. Worrall; Mr B.L. Duddridge; Mr Roberts; Geoff Evans; David Male; Eric Harlin; Margaret Fry; Mr E. Draycott; George Jenks and Mr L. Callaway (ex-works manager of George Parnall & Co.).

Readers may be interested in a book entitled *Parnall's in Memoriam*, (£2.50, published by Yate District Oral History Project), available from Yate Library, 44 West Walk, Yate, Bristol, BS17 4AX.

CHAPTER 1
Historical Survey

THREE PIONEERS

Before a more detailed study of Gloucestershire's 'big three' in aviation is made (Gloster, Bristol, Parnall), we should take a brief look at some pioneering work undertaken in the county by other companies and individuals in the field of aeronautics.

A prime example was the Webb-Peet tandem monoplane, built in 1910 by Webb, Peet & Co., Aeronautical Engineers, of Westgate Ironworks, Gloucester. This monoplane, designed by the brothers Scott, carried a pilot and passenger in tandem. The main advantage claimed was the location of the occupants and engine amidships, it being calculated that any difference in relative weights of the occupants would not affect the machine's balance.

The Webb-Peet monoplane had a small wing (canard) mounted at the front of the fuselage acting as an elevator, this front plane pivoting about the centre of pressure for ease of movement. The mainplanes incorporated a wing warping mechanism for lateral control. It was claimed the gull-shaped wing design, with front and rear planes having flexible trailing edges, would prevent the aeroplane heeling at a dangerous angle when turning. A pair of small balanced rudders, used for 'steering', was located at the tail-end of the fuselage.

Although it never flew, this canard wing monoplane by Webb, Peet & Co. of Gloucester was an innovative design. Visible are the tandem seats, chain-driven propellers and the twin-balanced rudders. A tricycle landing gear was fitted later. (Courtesy Clive Dodge)

The centre-fuselage-mounted Webb-Peet rotary engine drove a pair of tractor propellers, one to each wing, driven by an endless chain which, if broken, would put both propellers out of action. In theory the monoplane would then glide to earth. To neutralize gyroscopic effects both propellers revolved in the same direction, but opposite to that of the rotary engine. Three wheels were fitted to the machine, two small ones at the front and a single large one aft. On take-off, the two smaller wheels were lifted by manipulation of the elevator, and the full weight and balance was transferred to the main wheel just prior to lift-off. How Webb-Peet's monoplane would have behaved is a matter for conjecture as it never left the ground!

Another Gloucestershire engineer, Francis James Dodge, moved from Granville Street, Cheltenham, to Key Street, Gloucester, and with a silent partner formed the Phoenix Radial Rotary Motor Co. Known as Vulcan Works the business advertised as High Class Motor Engineers, and Specialists in Aviation Motors. In 1910 the famous airman S.F. Cody visited Gloucester to purchase a Phoenix rotary engine for his 'Flying Bus', paying a cheque to the Phoenix Co. for £138 15s. Six Phoenix rotaries, despatched from Gloucester to the USA, were lost when the ship carrying them sank en route. Unfortunately these engines were not insured, and when Frank Dodge's silent partner withdrew his money from the company, this promising Gloucester aero-engine business was forced to close.

The Phoenix air-cooled rotary itself was a 5- or 7-cylinder type with fuel and lubrication fed via the crankshaft. One unit (No. 6) was bench-tested as a twin with two engines running together by means of a double chain drive. Frank Dodge was obliged to take one or two French engine manufacturers to court for infringement of his rotary engine patent. He joined the RNAS in the First World War, but after 72 hours was sent home. His work on rotary engines was considered far more important to the country's war effort than serving in the RNAS!

An interesting story was related to the author by Cyril Palmer of Cheltenham, concerning his elder brother Frank, who served with the RFC during the First World War and became a keen aircraft enthusiast. After the war Mr Palmer senior owned a second-hand shop and acquired

Vulcan works, Gloucester, of the Phoenix rotary aero-engine company in pre-First World War days. Note the advert for Aero & Motor Boat at Olympia on left. (Courtesy Clive Dodge)

Phoenix twin rotary aero-engine, with double chain drive, in test rig at the Vulcan works, Gloucester, c. 1912–13. (Courtesy Clive Dodge)

anything worth selling. In 1920 he attended an auction sale at the Winter Gardens Pavilion, Cheltenham, where the Gloucestershire Aircraft Co. Ltd (GAC) had built and stored aeroplanes during and immediately after the war. Numerous aeronautical bits and pieces were on sale, and Mr Palmer bought a large quantity at a knock-down price. Surprisingly, on further examination the new acquisition included enough parts for three complete French Caudron biplanes plus some Gnome engines minus magnetos.

At the time Peugeot were offering a prize of £1,000 for the first person to perform a figure of eight flight in a 'man-propelled aeroplane'. Young Frank Palmer entered the competition utilizing Caudron components for a 15 ft span biplane. Parts from a Lloyd 'chainless bicycle' provided the pedals, and a shaft drive, pointed to the front of the machine, was fitted with a chain wheel, itself chain-geared to a small sprocket behind a wooden propeller made by H.H. Martyn of Cheltenham. Frank Palmer turned the propeller at 600 rpm with normal pedalling, and 1,000 rpm by hard pedalling. The total weight of the aircraft was 56 lb, but after poor weather conditions prevented attempts to fly the biplane, Frank converted it to a 30 ft-span monoplane. His first try at flying this version ended with wing-tip damage, and it was not until Good Friday 1925 that Frank was happy with the design. He made his initial flight in the repaired monoplane from Cleeve Hill, Cheltenham, and to achieve take-off speed asked a troop of Boy Scouts to tow him until the machine left the ground. Frank's brother was among the lads pulling on that day, and he confirms that the Frank Palmer monoplane took off and reached a height of some 15 ft with Frank pedalling very hard indeed. The monoplane remained aloft for 100 yards before landing, when it hit a rabbit hole and tipped onto its nose receiving serious damage. Subsequently, Frank Palmer designed gliders until 1940 and an account of his work appeared in *The History of Man-Powered Flight*, a book published two months after Frank died at the age of seventy-one. Percy Mills, who once worked on the *Gloucestershire Echo*, was interested in Frank Palmer's aviation activities, and several articles were apparently published on the subject in that newspaper.

Boy scouts hauling Frank Palmer's man-powered monoplane into the air on Cleeve Hill, Cheltenham, Good Friday 1925. This poor quality picture is included for its rarity. (Courtesy Cyril Palmer)

SUCCESS STORIES

Other aviation-connected enterprises, created and developed in Gloucestershire, became famous world-wide. Rotol Airscrews, for example, was established in May 1937 at Llanthony Road, Gloucester, settling in a year later in its new Staverton site to become a main supplier of propellers for many types of British military and civil aircraft. During the Second World War Rotol had its own Flight Test Department (FTD), and many types of aircraft were flown from Staverton airfield testing Rotol propellers. The author recalls an Armstrong Whitworth Whitley being an almost permanent resident at Staverton as it carried out its test-bed duties on Rotol research flying. A Vickers Wellington VI, the pressurized high-altitude variant (DR475), flew with Rotol FTD, and early Spitfire variants with three-blade Rotol airscrews were followed later by updated Spitfires with more powerful engines, some fitted with Rotol contra-rotating propellers. FTD test pilot Jack Hall was killed in a Spitfire IX with a contra-rotating propeller when his aircraft crashed from low altitude near Boddington on 29 July 1944. By the end of the Second World War Rotol had produced some 100,000 propellers, which had been fitted to more than sixty types of British military aircraft.

Prior to the Second World War aircraft alternators, generators, hydraulic pumps and so forth had been mounted on aircraft engines, but with the size and numbers of these accessories increasing some improvement was sought. Thus in 1940 Rotol started to design and develop a remotely mounted gearbox driven from the engine by means of a universally jointed open propeller shaft. In consequence accessories could be mounted and driven on the gearbox, but engine-driven items like fuel pumps and such like remained mounted on the engine. On 20 September 1945 the world's first flight by a turbo-prop aircraft was made at Church Broughton, when GAC test pilot, Eric Greenwood, flew Gloster Meteor F1 (EE227) powered by two Rolls-Royce RB50 Trent propeller-turbines with five-blade Rotol propellers of 7 ft 11 in diameter. These blades were later reduced to a little under 5 ft.

Post-war Rotol continued developing more advanced propellers and equipment for military and civil aircraft, one big success being the four-blade propeller for Rolls-Royce Dart turbo-props as fitted to Vickers Viscounts and other civil airliners. By 1960 more than 100 airlines and aircraft operators world-wide relied on Rotol products, which were used on aircraft such as the Fairchild F27 (licence-built Fokker Friendship), Grumman Gulfstream, Armstrong Whitworth Argosy and many other types. In development of the Fairey Rotodyne, the world's first vertical-take-off airliner ('killed off' by political expediency), Rotol provided two four-blade 'forward flight' propellers, two gearboxes for each of the Napier Eland turbo-prop engines, a constant speed unit and a throttle governor. It also produced bleed-air and direct-

Gloster Trent Meteor (EE227), with specially designed Rotol propellers. The world's first turbo-prop aircraft to fly on 20 September 1945 piloted by Eric Greenwood. (Rolls-Royce plc)

drive constant speed turbo-alternators, and an accessory gearbox for the Fairey Gannet anti-submarine aircraft, while another triumph was the eight-blade contra-rotating propeller fitted to the Westland Wyvern naval strike fighter, which operated during the Suez Campaign.

Rotol's FTD left Staverton in June 1947 to operate from Moreton Valence alongside Gloster Aircraft; the runway there was longer and allowed FTD to fly later marks of Spitfire and heavier Hawker Sea Fury naval fighter with greater safety. Rotol's FTD finally closed down during 1954. At the end of 1958 Rotol became part of the Dowty Group, and on 1 April 1960 the revised title Dowty Rotol Ltd encompassed Rotol's subsidiary, British Messier, and Dowty Equipment. Work performed at the time of writing includes regular overhaul of propellers from the RAF's Lockheed Hercules C1 Series transport fleet, plus those of military aircraft from Algeria, Oman and Sweden. The overhaul of non-Dowty/Rotol propellers from civil aircraft is also undertaken, for example the Hamilton Standard type used on turbo-prop airliners. Today Dowty Aerospace Aviation Services (part of the TI Group) remains in the vanguard of British aviation and will be ready for the challenge of the twenty-first century.

Dowty Group of course recalls another of Gloucestershire's famous sons, George Dowty, who was knighted in 1956. Dowty was on the design staff of the Gloster Aircraft Company in the early 1930s and came up with an idea to incorporate shock absorbers and brakes within an aircraft wheel. This internally sprung design featured a one-piece casting containing the brake and a centrally placed suspension system, with steel springs surrounding a nickel chrome forging which ran in four bronze slides. A Palmer brake and carrier plate was rigidly bolted to a flange on the wheel hub. Vertical travel allowance for the cantilever landing gear strut was 6 in, calculated as sufficient to absorb the landing shock of a small aeroplane.

Gloster's produced three of Dowty's experimental wheels, and an order for six wheels came from Kawasaki of Japan. But GAC misquoted their price for the order, and Dowty resigned in order to make the wheels himself. With very little capital, and the help of two mechanically minded friends, J.G. Bowstead and J.R. Dexter, George Dowty assembled the wheels at 10 Lansdown Terrace Lane, Cheltenham, for which he paid half-a-crown a week. A workbench, a hand-operated pillar drill and some hand tools served as equipment, but by the end of

Westland Wyvern S4 (VZ748), showing to advantage the type's eight-blade Rotol contra-rotating propeller. The retractable landing gear is by Dowty. (Westland Helicopters Ltd)

September 1931 the six wheels were en route to Japan aboard the steamship *Yasukuni Maru*.

At the end of October 1931 Dowty moved to a small workshop in Grosvenor Place South, Cheltenham, with Bowstead and Dexter as his first employees and Dowty Aircraft Components was formed. An agreement with Kawasaki for their licenced production of the internally sprung wheel, and his design for a shock absorbing strut with compression rubbers substituted by steel springs, kept the wolf from Dowty's door, but only just. Despite orders for his shock absorber struts, internally sprung wheels, tail wheels and some complete landing gears from several British aircraft manufacturers, as well as export orders from Denmark, Germany, Italy, Siam and Sweden, Dowty's financial status remained precarious until 1934, when he received a production contract for Dowty landing gear to be fitted to Gloster Gauntlet fighters for the RAF.

When Gloster's SS37 fighter (the prototype of the Gladiator) appeared it featured a new style Dowty landing gear in which a single bent tube of streamline section, mounted rigidly to the fuselage on each side, pointed diagonally downwards and outwards. Made from high tensile steel, these legs were stiff enough to absorb all landing and take-off loads, and took the brake load in torsion. Palmer pneumatic brakes and Dunlop tyres were used in this design, the main advantage of which was its drag reducing properties and ease of maintenance. Another Dowty development was to fit a small shock absorber unit down by the landing gear wheels. By a system of leverages small travel in the shock absorber itself was increased many times, yet the design was quite simple, could be mounted on a cantilever strut and, when suitably faired, resulted in a very clean landing gear. Because there was only a single member involved, it could also be incorporated easily into a retractable design.

Dowty excelled in the field of retractable landing gear and his design for the Heston Phoenix was the first installation of its kind in a British high-wing monoplane. This hydraulically operated, inward retracting landing gear featured Dowty's own patented 'Nutcracker' strut in which two opposed hydraulic jacks operated on an offset link at the centre of a side bracing tube divided by a knuckle joint. The jacks pushed the knuckle joint upwards, and consequent shortening of the tube pulled the compression leg and wheel inwards. Dowty also designed a forward retracting landing gear with the landing gear structure conforming to normal practice, the side 'V' comprising front shock strut and rear brace pinned to the front and rear spars respectively. The upper part of the

Gloster SS37 (prototype Gladiator) fitted with Dowty cantilever landing gear struts and internally sprung wheels. The pilot is P.E.G. Sayer. (GAC via Hawker Siddeley)

Dowty's fitted its 'nutcracker'-type retractable landing gear to the high-wing Heston Phoenix monoplane, an example of which is seen here with wheels retracted. (MAP)

shock strut was housed in a cylinder pivoted to the rear face of the front spar, this cylinder acting as a jack, with the upper part of the shock strut forming the piston, or ram. Retraction took place when oil was pumped below the piston, thus shortening the shock strut and lifting the wheel.

In 1935 Dowty Aircraft Components moved to Arle Court, a Gothic-style mansion house near Cheltenham, with 100 acres of land suitable for development. The house became Dowty's HQ and a factory was built at the rear. By 1939, as Britain went to war, Dowty had developed and perfected its hydraulic control systems for aircraft, and a year later became Dowty Equipment Ltd. By the end of the Second World War the company had produced some 87,000 landing gears and nearly one million hydraulic units. Since then over several decades Dowty's has produced wheels,

Seen here at Staverton in the 1960s, Smiths (Aviation Div.) trio of test aircraft. Left to right: de Havilland Dove, Hawker Siddeley 748 and Vickers Varsity. (Smiths Industries)

fixed and retractable landing gears, tail wheels, aircraft ski gear, hydraulic systems, high-pressure power and hand-operated pumps, aircraft control equipment and electric warning devices, switches, buzzers and indicators. By the end of the 1950s Sir George Dowty had built an industrial empire, and his enterprise is evident today in many fields of engineering on an international scale.

Smiths Industries at Bishops Cleeve have long been associated with the instrumentation and high technology involved in aviation and aerospace products. In post-war years its work with advanced landing techniques, blind approach systems and other high-tech avionics, warranted the company using a fleet of three test aircraft at Staverton, a de Havilland Dove, Vickers Varsity and a Hawker Siddeley 748. As Smiths Aviation Division, it took over the airport from 1956 when Cambrian Airways moved out, and provided a manager, an airport fire service and an air traffic control until Cheltenham and Gloucester took over in 1962. Smiths first Varsity (G-APAZ) unfortunately crashed on a house in Tuffley on 27 March 1963, killing two crew members. A second Varsity (G-ARFP) was acquired as a replacement, eventually returning to the RAF as WF387 in May 1968. Numerous smaller Gloucestershire companies, deserving of a book to themselves, provided back-up to both local and national aircraft manufacturers, and were scattered far and wide around Gloucester, Cheltenham, Stroud, Stonehouse, Dursley, north Bristol, Cirencester, Tewkesbury and other areas of the county. Such firms were the backbone of Britain's wartime aviation industry.

AIRFIELDS

Many airfields, some permanent and famous, others known only briefly before passing into obscurity, have played an important part in Gloucestershire's aviation history. Perhaps the best-known is Staverton, Cheltenham and Gloucester's municipal airport, which has survived as both a military and civil concern in war and peace.

Origins of Staverton airport can be traced to an airstrip owned by the Cotswold Aero Club at Down Hatherley from 1932. Plans for a public aerodrome could not be met on this site and proposals for an airfield on the other side of the A40 main road, between Gloucester and Cheltenham, were accepted by both councils. In March 1934 they jointly purchased 160 acres at the new location, work starting the following November. Two years later, on 18 November

1936, an operating licence was granted, the first scheduled service to use Staverton being the Bristol to Birmingham route of Railway Air Services.

By 1937 an agreement with the Air Ministry decreed that in return for improvements to the aerodrome it could be used for a period of twelve years for training RAF aircrew. Night flying equipment was installed and new buildings erected for the use of a reserve flying school. At the end of September 1938 No. 31 E&RFTS formed with DH Tiger Moths, Airwork Civil School of Air Navigation (later 6 Civil Air Navigation School) arriving the following May with DH Dragon Rapides to train RAF observers. At the outbreak of the Second World War 31 E&RFTS closed as the Reserve aspect was dropped to conform with new RAF training schedules. During September 1939 the airport became RAF Staverton, and Airwork's units became 6 Air Observer Navigation School (AONS). The Dragon Rapides were replaced by Avro Ansons, and from August 1940 more Tiger Moths joined the circuit after two EFTS arrived on site. During 1941 AONS detached flights to Moreton Valence and Llanbedr, and in January 1942 became No. 6 Air Observer School (later 6 (O) Advanced Flying Unit). Another RAF presence at Staverton was 44 Group's Communication Flight, which closed down in August 1946.

Meanwhile, at the end of 1940 Folland Aircraft's 43/37 test-beds arrived at Staverton for trials with Bristol and Napier engines and Rotol propellers. One of these aircraft broke up in mid-air while on a diving test with a Bristol Hercules engine, GAC test pilot Michael Daunt managing to escape by parachute. GAC moved part of their FTD to Staverton and the author recalls seeing a long line of Hawker Typhoons lined up on the airfield.

Dowty's use of Staverton as a test centre included landing gear trials on a four-engined Halifax bomber. This aircraft caused quite a stir among the locals as it made numerous low circuits of the airfield during these tests.

Flight Refuelling was based at Staverton from 1942 until 1946, and as well as developing its in-flight refuelling techniques this company carried out tests with anti-icing devices, flame-damping exhaust trials, self-sealing fuel tank experiments and towed a Hawker Hurricane behind a Wellington bomber as a possible means of ferrying fighters to Malta. Flight Refuelling

A snowy day at Staverton, March 1942, and a Bristol Hercules VIII is about to be tested in Folland 43/37 engine test-bed, P1775. Note the sturdy fixed landing gear. (Rolls-Royce plc)

left Staverton in 1946, leaving Rotol and a few civil flyers as sole occupants until the RAF Police Guard Dog Training School arrived, followed by No. 1 RAF Police Wing.

The end of September 1950 saw the airport back in civil hands, but not under a particular authority. Cambrian Airways moved in during 1953, but found Staverton unprofitable and withdrew in 1956. Smiths Aviation Division remained until 1962, when the airport passed to Cheltenham and Gloucester councils as a municipal airport. Other concerns using Staverton have included Skyfame Museum, Glosair (UK agents for the Victa Air-tourer), Derby Airways, British Midland, Intra Airways and Dan-Air which flew the Jersey route with Hawker Siddeley 748s. Staverton airport has also been home to Flight One and several companies involved in private charter and light cargo flights, aircraft maintenance and sales, and of course several flying schools and clubs, including Cotswold Aero Club, Gloucester and Cheltenham School of Flying and Staverton Flying School.

Aston Down airfield, near Chalford, survived as a military concern for quite a long period. It opened on 12 October 1938 as No. 7 Aircraft Storage Unit, soon to become 20 MU. In August 1939 RAF Fighter Command moved in to provide pilots with intermediate training on Harvards, Gladiators and Blenheims. This became 5 OTU in March 1940 as Spitfires, Hurricanes, Defiants, Masters and Fairey Battles arrived, and from November 1940 to March 1941, 55 OTU used Aston Down for training Hurricane and Blenheim night fighter pilots.

Ferry flights from Aston Down were undertaken by the ATA, many of whose pilots were women, while following units included 52 OTU (Masters and Spitfires), FLS (Spitfire VBs), 1311 Flt (Anson ambulances), 3 Tactical Exercise Unit (rocket-firing Hurricanes and Typhoons), which at the end of 1944 became 55 OTU (Hurricanes, Typhoons, Martinets and Masters) before disbanding in June 1945. By July over 1,000 aircraft were crammed into Aston Down in charge of 20 MU, while the ATA unit gave way to a glider tug Flight with Whitleys and Albemarles, as well as Harvards and Oxfords from 41 Group Training Flight.

Aston Down gradually reduced its stock of stored aircraft, and after 20 MU closed down in 1960 the airfield was used by the CFS, Little Rissington, as a relief aerodrome until 1976. The Ministry of Aviation, responsible for storage space at Aston Down since 1963, has maintained the site in excellent condition, and for a number of years the airfield has been used by the Cotswold Gliding Club, which launches its gliders from either a ground winch or aircraft tug.

One of the RAF's well-known Gloucestershire bases was Little Rissington, a Cotswold airfield continuously in use from the start of the Second World War until the late 1970s. No. 6 FTS arrived in August 1938, becoming 6 SFTS when war broke out. Aircraft then flying from Little Rissington included Audaxes, Harts, Furies, Harvards, Ansons and, later, Oxfords.

German air activity over the county increased during 1940, and on the night of 18 August an Anson from Little Rissington, flown by Sergeant B. Hancock, collided with a Heinkel He 111 bomber. Both aircraft crashed killing everyone on board, and it has remained a matter for conjecture whether or not Sergeant Hancock deliberately rammed the enemy aircraft. If he did, it was an act of great courage and self-sacrifice.

By 1942 pilots who were used to flying overseas had to acclimatize themselves to UK conditions. Navigation, night flying and beam approach training was carried out at Little Rissington, while in another part of the base 8 MU prepared aircraft for service throughout the war. This unit closed down in 1957 after one of its main tasks – scrapping large numbers of redundant Vickers Wellington bombers – had been completed.

Shortly after the Second World War the CFS reformed at Little Rissington, training beginning initially on Harvard two-seat trainers. After the dreary years of war the public welcomed ceremonies and they certainly had their fill from the CFS on 15 July 1953, when a formation of twenty-three Harvards from the CFS flew in the fly-past for HM The Queen's Coronation Review.

By then many RAF piston-engined aircraft were being replaced by Meteor T7 jet trainers at

Little Rissington. In 1954 Vampire TIIs began supplanting Meteor T7s, then came the Jet Provosts. One still observed an odd Chipmunk, Prentice, piston-Provost, and of course Varsity and Anson C19/21 twin-engined types, well into the 1960s, but time was running out for Little Rissington as a major RAF base even after the agile Folland Gnats arrived in July 1962. Eventually the MoD closed the station for RAF purposes, the CFS moving initially to Cranwell. Since then any flying at the base has mainly involved 637 ATC Volunteer Gliding School in powered Slingsby Ventures.

Over a number of years RAF Fairford has become famous as the bi-annual venue for the International Air Tattoo. However, fifty years ago the base was home to 190 and 620 Squadrons, RAF, flying Short Stirlings, these being followed by Halifaxes of 47 Squadron which remained at the base for two years. During the 1950s Fairford was home to USAF B-47/RB-47 Stratojets, and in 1966 the RAF's 53 Squadron reformed there with its massive Short Belfast heavy transports.

In 1969 and the early 1970s Fairford provided a base and Flight Test Centre for Concorde, the supersonic airliner. This attracted many people from near and far who watched in awe as this magnificent example of Anglo-French technology and co-operation landed and took off from Fairford's runway as it underwent various trials. Also during 1969, 30 Squadron was reformed at Fairford with Lockheed C-130K Hercules transports, this unit moving to RAF Lyneham in 1971 as part of the LTW.

The late 1970s and 1980s saw Fairford hosting the long-range USAF Boeing KC-135 in-flight refuelling tankers, which flew to all parts of the compass, replenishing in mid-air the fuel tanks of USAF (or Western Alliance) military aircraft thus greatly extending their range and duration. More recently Fairford was used operationally, almost on a war footing in fact, as it provided a base for USAF Boeing B-52 heavy bombers flying vital missions during the 1990/91 Gulf War.

Filton airfield (now in Avon) is in reality a part of the Bristol Aeroplane Company, and has remained so for over eight decades. Therefore its main story will be found in chapter 3 which covers the Bristol company.

RAF Kemble, near Cirencester, was home until the 1980s to 5 MU. The oldest maintenance unit in the RAF, its work over forty-five years has been constant in handling a large selection of aircraft

Red Arrows Gnats break formation over a Red Pelican team Jet Provost T5 (XW288) at Little Rissington in 1974. (Courtesy of the Gloucestershire Echo)

USAF Boeing KC-135A Stratotanker at Fairford, 1980. (MAP)

Battle of Britain Memorial Flight Lancaster and Concorde 002 at Fairford, c. 1969. (Courtesy Mr & Mrs A.R. Hacker)

types. Starting as 1 ASU at Peterborough in 1932, it was responsible for the storage and issue of aircraft. Moving to Waddington as 'H' MU in 1936, its growth steadily increased until, by the time it moved to Kemble as 5 MU in January 1939, it was dealing with all types of aircraft. During the war these included Hinds, Whitleys, Blenheims, Wellingtons, Spitfires, Hurricanes, Lancasters, Beauforts, Albemarles and Typhoons. A Tug and Glider Flight was also at Kemble to test Horsa and Hotspur gliders assembled on site. It also hosted an Overseas Aircraft Delivery Squadron, and over 500 Wellington bombers were modified and delivered to RAF Middle East units.

During 1943 runway extensions and new taxiways were built to connect the many dispersal points around Kemble, some of which were 1½ miles out! A vast number of Hawker Typhoons were prepared in readiness for D-Day, and when this task finished in the summer of 1944 Lancaster bombers began arriving to be re-engined with Rolls-Royce Merlin 24s. In January 1945 Kemble was virtually taken over by 47 Group, RAF Transport Command, with 5 MU a mere lodger. By the end of 1945 some 1,030 aircraft were on charge to 5 MU, but with the selling off of surplus machines that had been impressed into service, and with the scrapping of many airframes, numbers soon dwindled. After 47 Group had withdrawn from Kemble in 1946, 5 MU quickly became one of the most important in the UK.

The first DH Vampire jets arrived in 1946, and later 5 MU was responsible for modifying 54 Squadron's Vampire F3s prior to their becoming the first RAF jets to cross the Atlantic in July 1948. During 1952 batches of Canadair CL-13 Sabre fighters for RAF and 2nd Tactical Air Force service were received to be modified and painted in appropriate RAF camouflage finish. Some 550 of these Sabres passed through Kemble's shops in this operation, code-named 'Beechers Brook'. Two years later Hawker's beautiful Hunter entered RAF service, and from that time 5 MU was heavily involved in modernization programmes for RAF and Royal Navy Hunters.

The 1960s saw 5 MU servicing and updating many aircraft types such as Andovers, Jet Provosts, Canberras, Shackletons, Sea Princes, Varsitys and Chipmunks. In 1962 Kemble's new Air Traffic Control Tower was opened, and in 1964 5 MU celebrated its 25th anniversary with an impressive open day attended by over 20,000 people. The RAF's famous Red Arrows aerobatic team, with its bright red Folland Gnats, was based at Kemble and they remained there until the 1980 season when BAe Hawks replaced the Gnats. After the Red Arrows moved to Cranwell, the 1980s saw the demise of Kemble as an RAF base and airfield, the USAF moving in to make it their Logistics Command Support Centre (Europe).

Details of Brockworth/Hucclecote and Moreton Valence airfields will be found in chapter 2, as they were very much an integral part of the Gloster Aircraft Co. from its early days until its demise in the early 1960s. Yate aerodrome was similarly an intricate part of George Parnall & Co., as well as Parnall Aircraft Ltd later, and it too is incorporated into those companies' stories in chapter 4.

Over the years Gloucestershire has provided the RAF with numerous bases, like South Cerney which was home to 3 FTS as early as 1938 (3 (P) AFU in the Second World War), and in post-war years remained a training base well into the 1960s, its piston-Provosts being a familiar sight over the Cotswolds. But other airfields were a wartime measure – Broadwell, Down Ampney and Blakehill Farm for example – from where Dakotas of 48, 233, 271, 512 and 575 Squadrons carried Allied paratroops and towed Horsa heavy transport gliders to Normandy on D-Day, 6 June 1944.

Babdown Farm, near Tetbury, was used for RAF trainers on night flying sorties, Harts and

Some 550 Canadair F-86E Sabre jets were 'shopped' for the RAF at 5 MU, Kemble, in the 1950s. These three are with 234 Squadron, 2nd TAF, Germany. (Author's collection)

This Hunting Percival Provost T1 (XF841/'03') of the CFS, was a type flown regularly from South Cerney. It is seen here at Little Rissington in 1964. (MAP)

On D-Day 1944 many Horsa transport gliders, like this one, were towed from several Cotswold airfields, and later on to the operations at Arnhem. Note the D-Day invasion stripes. (MAP)

Audaxes from 9 FTS, followed (after new runways, hangars, HQ site and training huts had been added in 1942) by Oxfords and Masters from 3 FIS, and Spitfires of 52 OTU. In 1943 a standard beam approach was fitted for the main runway and used by Oxfords of 1532 Blind Approach Training Flight. No. 15 (P) AFU moved to Babdown in 1944, and its last aircraft left the airfield on 20 June 1945.

Barnsley Park, near Cirencester, opened in June 1941 as a satellite airfield, but proved difficult for flying purposes. Three Hurricanes arrived there in November 1943, and numerous aircraft used the site during 1944 as a relief to Kemble, but by October 1945 Barnsley Park was

A number of Australian pilots of the AFC were killed and injured while flying Sopwith Camels, like this one, from Leighterton during the First World War. (Author's collection)

closed. Down Farm, another satellite airfield near Tetbury, opened in April 1941, its early dispersals including Defiant two-seat turret fighters, Hampden bombers and Oxfords from 15 (P) AFU doing 'circuits and bumps'. By January 1946 Down Farm had been vacated.

Chedworth airfield was opened in 1942, and between then and the end of 1945 was used by Spitfires of 52 OTU and the FLS, Oxfords of 3 and 6 (P) AFU, Mosquitoes and Martinets from 60 and 63 OTUs, Stinson L-5 Sentinels of the USAAF (9th Air Force HQ and 125th Liaison Squadrons) on communication duties, Mustangs from C Squadron of 3 TEU and Typhoons from 55 OTU. CFS also utilized Chedworth for forced landing exercises in post-war years, but by the end of the 1970s the airfield had reverted to agricultural use.

Other RAF satellite airfields in Gloucestershire included Long Newnton, near Kemble, used by 3, 14 and 15 SFTSs and 3 (P) AFU; North Stoke (Bath's Lansdown racecourse converted for flying and used by 3 FIS from 1943 until July 1945); Overley, near Cirencester, used mainly as a dispersal area for various types of aircraft, including four-engined types, and also as a relief landing ground during 1942 for the Oxfords of 3 SFTS.

Small satellite airfields were also in use at Bibury and Windrush, and just north of Cheltenham was Stoke Orchard aerodrome, from where Miles Masters towed Hotspur training gliders aloft. Many glider pilots who flew in the D-Day operations of June 1944 were members of the Glider Pilot Regiment, having done their training at Stoke Orchard.

In the First World War pilots from the Australian Flying Corps (AFC) flew Avro 504s and Sopwith Camels from Leighterton, and 6 (T) Squadron, AFC, used Minchinhampton airfield flying Bristol Scouts, Sopwith Pups and 1½ Strutters. These were later joined by 5 (T) Squadron with Farman Shorthorns, DH 6s and more Sopwith Pups. Eventually 6 (T) Squadron was issued with SE5a single-seaters, and by the autumn of 1918 there were twenty-four Avro 504Ks, twelve Sopwith Camels and twelve SE5as on site. After the Armistice flying practically stopped, the Australians leaving in January 1919. They left twenty-four of their fellow pilots behind, buried in Leighterton cemetery, most having been killed in Sopwith Camels. Minchinhampton quickly returned to agricultural use, but eleven years later work began on its reversion to an airfield named Aston Down.

CHAPTER 2

Gloster Aircraft

The origins of the Gloster Aircraft Company can be traced back to 1910. Aeroplanes were in their infancy, and young George Holt Thomas, whose father owned the *Daily Graphic* newspaper, saw flying machines as big news. He became obsessed by aeroplanes and flying, produced a number of airships and kite balloons, and in 1912 formed the Aircraft Manufacturing Co. (Airco) at Hendon. During 1913 young Hugh Burroughes joined the firm, a man who would later become a cornerstone in developing Gloster Aircraft. He had worked at Farnborough in the Balloon Factory, later Royal Aircraft Factory (R.A.F.), where he first met Holt Thomas. Within a year Burroughes was manager at Airco, and had been joined by Geoffrey de Havilland as chief designer.

The outbreak of the First World War in August 1914 increased the demand for aeroplanes, and obviously Airco would need to subcontract their work. As aircraft were then of mainly wooden construction, it was logical to approach a firm with expertise in cabinet making and wood products. Hugh Burroughes, with Airco's production manager, Guy Peck, chose H.H. Martyn & Co., Sunningend works, Cheltenham. An agreement was signed between Airco and Martyn's, the first contracts being to produce parts for Farman Longhorn and Shorthorn biplanes (a manufacturing licence having been obtained for these French biplanes by Holt Thomas during a visit to the Farman brothers in France).

In 1917 it was proposed a new company be formed with Martyn's and Airco having equal 50 per cent shares. It was agreed the newly formed company would rent Sunningend works from Martyn's, and take over Airco's subcontract work. This arrangement was beneficial to all parties, and on 5 June 1917 Gloucestershire Aircraft Co. Ltd (GAC) was formed and officially registered.

By the end of 1917 GAC had built 150 Airco DH6 biplanes, followed in 1918 by 150 Bristol F2B fighters. From a second order for 250 Bristol F2Bs, only 135 machines were built as part of the contract was cancelled. Even so, by the middle of 1918 GAC was so inundated with work it had to subcontract work out to other local firms like Savages Ltd, Daniels & Co., and the Gloucester Carriage and Wagon Co.

Meanwhile in 1915 the Air Board had established an Aircraft Acceptance Park and Mobilization Station near the village of Hucclecote. GAC transported its aircraft there towed behind a lorry, the wings having been removed and carried separately. Specially adapted road wheels were fitted to the aircraft for this journey between Cheltenham and Hucclecote via Shurdington, a practice that would continue until GAC moved to Hucclecote during the 1920s. The Oddfellows Inn at Shurdington, about half-way between, was well known as a stopping place where the lorry crew allowed the aircraft's axle to cool off! It was also a regular stop for the wages van after its deliveries to GAC's sites. By 1918 the Hucclecote/Brockworth airfield covered 140 acres, with 20 acres of buildings including five hangars, 21 storage sheds, a transport depot and an ammunition store. The RAF's 90 Squadron was based there for a short time in July 1918 with Sopwith Dolphins, but quickly disbanded. GAC rented hangar space at Hucclecote in 1921, but within seven years it had purchased the entire site.

The Nieuport Nighthawk, an aircraft produced by GAC at the end of the First World War, was a single-seat fighter designed by Henry Folland, who became chief designer at GAC for well over a decade. Folland had been responsible for designing the SE5a fighter before joining

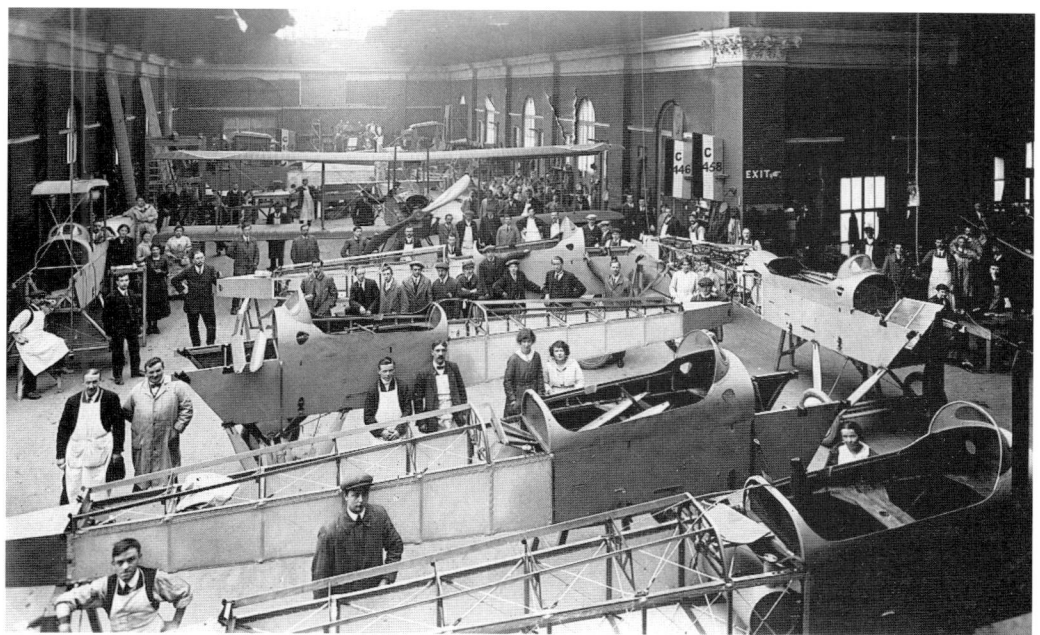

In 1917 the Winter Gardens Pavilion, Cheltenham, was used by the Gloucestershire Aircraft Co. for the production of de Havilland DH6 biplanes. (Author's collections)

Chief designer and engineer for Gloster's for some fifteen years was H.P. Folland, seen here without his pipe! (GAC)

Single-seat Gloster Sparrowhawk I (JN400) built for the Imperial Japanese Navy. (GAC)

Nieuport & General (British Nieuport) at Cricklewood as chief designer. His Nighthawk of 1918 was powered by a 320 hp ABC Dragonfly, an engine which proved a failure because it was overweight, underpowered and prone to severe vibration. It led to the collapse of Nighthawk production, but Folland converted his Nighthawk design into the LC1 Nieuhawk racer, which flew in the 1919 Aerial Derby. A year later J.H. 'Jimmie' James (later GAC test pilot) took part in the same race flying a second LC1. However, Folland's Nieuport Goshawk racer crashed, killing the famous airman Harry Hawker, and British Nieuport closed down in November 1920. GAC snapped up design rights for the Nighthawk fighter, also acquiring the services of Folland as consultant engineer and later chief designer. Folland's first task was to convert his Nighthawk design into the Gloster Mars II, III and IV series, which were built for the Japanese Navy as the Sparrowhawk I, II and III respectively. These machines, similar to the Nighthawk, were powered by a 230 hp Bentley BR2 rotary engine. Mars VI was basically a Nighthawk with metal fuselage components and an Armstrong Siddeley Jaguar engine. Mars X (named Nightjar) was a shipboard naval development with a Bentley BR2, and flew with 203 Squadron (Leuchars) and 401 Flight aboard the aircraft-carrier HMS *Argus*. In 1922 six of 203 Squadron's Nightjars embarked aboard *Argus* in readiness for the Chanak crisis.

MARS I (BAMEL)

Folland, determined to put GAC in the forefront with fast military aircraft, developed the Nighthawk into an advanced racing machine. He was convinced Air Ministry encouragement would eventually be forthcoming to firms capable of building exceptionally fast aircraft. This new racer, a single-bay biplane designated Mars I (named Bamel), was produced when the home market for military aircraft was practically nil, but Folland undoubtedly achieved his ambition for two years later the Gloster Grebe fighter entered RAF service. Gloster's Bamel (G-EAXZ) was built at Sunningend and made its first flight from Hucclecote on 20 June 1921.

It featured a streamlined engine cowling surrounding a 450 hp Napier Lion engine, streamlined propeller spinner, a retractable radiator and single I-type spruce interplane struts. To squeeze the Lion engine into his racer's small dimensions (wingspan 22 ft, length 23 ft) Folland maintained balance but sacrificed forward visibility by mounting the fuel tank behind a finned water header tank, both within a faired humplike cabane supporting the upper wing. This hump (Camel) and the type's sturdy build (Bear), half Bear-half Camel, is reputed to have given this racer the name Bamel. It was entered in the annual Aerial Derby at Hendon on 17 July 1921, but four days before the race it was damaged in a landing incident. Quick repairs had the Bamel ready in time for the race, but en route to Hendon the aluminium carburettor intake pipes cracked. These were replaced at the last minute with steel pipes provided by de Havilland's at Stag Lane, who sent them via a fast motor cycle to Hendon. In the end Jimmie James flew Gloster's racer to victory, winning at an average speed of 163.3 mph.

GAC's team, now inspired to enter the Bamel in France's Coupe Deutsch Race at Villesauvage, reduced the racer's wing area and replaced the square type radiator with twin Lamblin 'lobster pot' radiators. Unfortunately James was forced to abort from the race when fabric began bulging along the wings. On its return to Sunningend the Bamel was modified with a further reduction in wing area, a revised method of attaching the fabric, new sheet metal fairings at the I-strut corners, over cylinder heads and immediately aft of the tyres, and installation of a single Lamblin radiator.

On 19 December 1921 James flew the Bamel to Martlesham for an attempt on the world air speed record, but although reaching a speed of 212 mph was just short of the necessary increase in speed to beat the then existing French record. The four passes over the course averaged 196.6 mph, the French world record then standing at 205.22 mph. The Bamel had, however, established a new British speed record of 196.4 mph.

For the 1922 Aerial Derby Folland made further modifications to G-EAXZ, with revised fin and rudder, deletion of the wheel fairings and reversion to twin Lamblin radiators. Once again James romped home a second-time winner at a speed of 177.85 mph. Also entered for the 1922 Coupe Deutsch Race, the Bamel would easily have won as all the other competitors except one (a

In original configuration at Hucclecote in 1921, the Gloster Mars I (Bamel), with 450 hp Napier Lion engine. (GAC)

J.H. 'Jimmie' James, Gloster's first test pilot, seen here in the cockpit of the Mars I (Bamel). (Courtesy of the Gloucester Citizen)

slower machine) were rendered *hors de combat*. Unfortunately, however, Jimmie James lost his maps from the open cockpit and became hopelessly lost over France. Ironically he flew the course next day faster than any other type there, which to Folland was a pointless gesture. Another attempt on the world air speed record in October 1922 again failed, and when his contract with GAC ended on 1 April 1923 Jimmie James was replaced as chief test pilot by Larry Carter.

Meanwhile, in November 1922 the Bamel had been flown at the A&AEE, Martlesham, by Flt Lt Rollo de Haga Haig, who pushed the racer skywards at nearly 3,500 ft/min, and at 5,000 ft the climb rate was 2,390 ft/min. At 19,500 ft the Bamel's climb was still remarkable at 854 ft/min, and for his efforts Flt Lt de Haga Haig won the Royal Aero Club Certificate of Performance.

For the 1923 Aerial Derby Folland fitted an uprated Napier Lion engine, relocated the fuel tank to the fuselage, altered vertical tail surfaces to include a horn-balanced rudder, and introduced N-style wing cabane struts. In this form the Bamel, or Mars I, was renamed Gloster I. Incentive was added to the 1923 race in an Air Ministry statement offering to purchase the winning aeroplane, less its engine, for £3,000.

In the event Gloster I, flown by Larry Carter, came in first at an average speed of 192.4 mph, and thus became Air Ministry property as J7234. Fitted with twin floats, Gloster I tested various floats for Schneider Trophy racing seaplanes, and in 1925 and 1927 it was used to train RAF pilots for the contest. During its time in RAF service several modifications were made to J7234, including a header tank on the upper wing and flush radiators in upper and lower wing surfaces. Gloster I was scrapped in 1927, but from her evolved a line of Gloster racing seaplanes as contenders for the coveted Schneider Trophy.

After transferring to the Air Ministry as J7234, the Bamel was modified to seaplane configuration as the Gloster I. It is seen here at Felixstowe, c. 1924. (GAC)

THE RACING SEAPLANES

Following the Bamel's success as a racer, and its adoption as a twin-float seaplane by the RAF High Speed Flight, Folland and GAC set their sights on the Schneider Trophy, a magnificent bronze statuette then valued at £1,000, to which was added a cash prize of £1,000 for the winner of three consecutive contests. However, without the generosity afforded by Lady Houston, who personally guaranteed an expenditure of £100,000 in the venture, England could not have entered the Schneider Trophy Contests (politicians were unwilling to give financial support despite Great Britain's prestige in the air being at stake).

At GAC Folland developed his Gloster II from the original Gloster I, the new machine having a 585 hp Napier Lion VA. To maximize streamlining and reduce drag, fairings were fitted around all three cylinder blocks, centre-plane struts faired in and radiators mounted on the float struts. Of two Gloster IIs built, the first (J7504) went to the MAEE, Felixstowe, on 12 September 1924 to be tested by Capt Hubert Broad. However, while the machine was alighting on rough water, a float support strut fractured causing the racer to sink and become a complete write-off. The Schneider Trophy Contest was in fact cancelled in 1924 and held over until 1925, a very sporting gesture by the Americans who could have had an undisputed win.

The second Gloster II (G-EBJZ) crashed at RAF Cranwell the following year in landplane configuration. At the time it was flight testing new equipment for Folland's new Gloster III racer and hit the ground during a fast run at low altitude. Test pilot Larry Carter was seriously injured in the crash and died fifteen months later in a Cheltenham nursing home.

Entered for the 1925 Schneider Trophy Contest Gloster III was arguably the most successful of Gloster's racing seaplanes. Folland, determined to persevere with a biplane configuration, despite R.J. Mitchell's adoption of a monoplane layout for his Supermarine S4, created the smallest aeroplane in relation to its power that had been built in the UK up to that time. Thus Gloster III emerged as a sleek 20 ft-span biplane powered by a 12-cylinder 700 hp Napier Lion VII driving a Fairey-Reed two-blade metal propeller. Construction was mostly wood, the fuselage being a plywood-covered monocoque, while the wings featured very thin aerofoils and were fabric-covered.

Flush-fitting wing radiators were intended, but these were not ready in time and the Lamblin type had to suffice. These oblong units, fitted as they were to the lower wings, spoiled the aircraft's appearance and induced more drag. After trials at Felixstowe in August 1925, the first Gloster III (N194) was found wanting in directional stability. This resulted in dorsal and ventral fin areas being increased and the aircraft redesignated Gloster IIIA. A second IIIA (N195) was prepared for the Schneider Trophy Contest, but flew only once at the MAEE before being shipped to Baltimore, USA, the venue of the great event on 24 October 1925.

During pre-contest trials at Baltimore, the Supermarine S4 crashed, leaving the two Gloster racers as Great Britain's only hope. That diminished when N195, piloted by Bert Hinkler, was badly damaged during taxying trials leaving Hubert Broad to fly alone for Great Britain. Lt James H. Doolittle, US Army, flying a Curtiss R3C2 racer took off first, with Broad in N194 following five minutes later, the rest of the field likewise taking off at five-minute intervals. These were the Americans Cuddihy and Ofstie in Curtiss Navy racers and the Italian de Briganti in a Macchi 33. Lt Doolittle won the contest with a speed of 232.5 mph, Broad was second, de Briganti third, while Cuddihy and Ofstie were forced to retire with engine trouble.

Broad did extremely well to bring the Gloster IIIA in second at a speed of 199.169 mph, having lapped at 200 mph, faster than any previous Schneider Trophy winner. Broad remarked that had it not been for the extra drag imposed by the Lamblin radiators and necessary wide pylon turns, N194 would probably have won as it was only 33 mph slower than Doolittle's Curtiss. Even so N194 set the British 100 kilometre record, and on returning to GAC was modified for use as a propeller test-bed.

Meanwhile, N195 was updated to have surface radiators on all four wings, a large expansion header tank in the upper centre-section, an enlarged windscreen, full cantilever tailplane, a

Gloster III racer under construction at GAC, Sunningend, Cheltenham, c. 1925. (GAC)

rounded leading edge to the fin and round bracing wires superseded by a streamlined type. As such, N195 was redesignated Gloster IIIB, and achieved 252 mph at sea level. It joined N194 at the RAF High Speed Flight, where both racers were used as development and training machines in helping Great Britain to prepare for the 1927 Schneider Trophy Contest in Venice.

For this event Folland still retained a biplane layout, his Gloster IV being a great advancement over its predecessors. It contained components producing higher aerodynamic efficiency to the extent of an extra 70 mph above the top speed of the earlier Gloster III racers. Again of mainly wooden construction, the Gloster IV featured a large fin and spine built integral to the fuselage, upper wing roots sweeping down to be faired aft of the Lion engine's top two cylinder banks and lower roots curved sharply upwards into the fuselage. For this racer Gloster designed its own Duralumin floats, which were attached to the fuselage by tubular steel struts contained in wooden fairings. Radiators were installed in the float tops and wing surfaces.

Three Gloster IVs were built, the IV (N224), IVA (N222) and IVB (N223). The IV and IVA were powered by a 900 hp direct-drive Napier Lion VIIA engine, the IVB having an 875 hp geared Napier Lion VIIB. On N222 and N223 wing area and span were reduced, and N222 was provided with cruciform-style vertical tail surfaces.

All three machines arrived at Calshot during the summer of 1927 to carry out initial flight tests before N222 and N223 went to Venice. The Schneider Trophy Contest started at 2.30 p.m. on 26 September, with Gloster IVB (N223) taking off piloted by Flt Lt S.M. Kinkead. Five-minute intervals elapsed between take-offs and each competitor was allowed ten minutes for taking off, making a circuit and gaining altitude prior to flying to the starting line. Kinkead was followed by an Italian Macchi M52 flown by di Bernardi, Supermarine S5 (N220) with Flt Lt S.N. Webster, another Macchi M52 piloted by Guazzetti, a second Supermarine S5 (N219) flown by Flt Lt O.E. Worsley, and a third Italian Macchi M52 in the hands of Ferrarin.

An exciting start to the contest saw di Bernardi and Kinkead practically neck and neck on the first lap, but the Italian was forced to withdraw at the end of the second with engine trouble after clocking 263.1 mph. Gloster's N223 flew on to increase its lap times, but during the fourth lap lost its propeller boss due to excessive vibration caused by loose propeller shaft splines, and Kinkead had to retire. All three Italian machines dropped out, leaving the two Supermarine S5s battling it out until Webster finished the winner at an average speed of 281.65 mph, with Worsley second at 273.01 mph. There was some consolation for Glosters as N223 had, in its third lap, set an all-time record for biplanes of 277.1 mph. The Gloster IVs were later modified, two acting as trainers to the High Speed Flight, while N224 was sold privately in 1930 when it was planned to use it in an attempt on the World Air Speed Record, an idea that never matured. N223 crashed trying to land in fog on 19 December 1930, but N222 remained with the High Speed Flight contributing much to the success of Britain's Schneider team, especially in their final triumphant win of 1931.

Gloster's final Schneider Trophy attempt was the beautiful Gloster VI 'Golden Arrow', a sleek monoplane of which two were built in 1929 for that year's contest. Folland had finally switched from his beloved biplane to produce this magnificent monoplane, but his loyalty to Napier was maintained with the installation of a 1,320 hp Lion VIID, a new engine which plagued Gloster's team throughout the contest.

The Gloster VI featured a semi-monocoque metal fuselage with Duralumin skinning, but the wings were made of wood with laminated wood covering. Thickness and maximum chord applied to the outer sections provided a thin wing with low drag (inner portion) plus a thick wing and high lift. Two Gloster-designed metal floats acted as main fuel tanks, the petrol being raised to a fuselage collector tank via engine-driven pumps. Radiators were flush-fitted to the wing surfaces, and an oil cooler was installed to fit flush round the fuselage aft of the cockpit. Further oil coolers were located in the twin floats upper surfaces, the oil circulating through

Gloster IV racing seaplane (N224) with broad-chord Gloster propeller and empennage adopted for IVA and IVB later. Seen here at Calshot, c. 1927. (BAe plc, Manchester)

Beautiful Gloster VI 'Golden Arrow' Schneider racer shows off its sleek lines in 1929. (GAC)

special pipes in the leading edge of the float struts.

Both Gloster VIs (N249 and N250) went to Calshot in August 1929 for trials, and were test flown by three RAF Schneider pilots, Sqn Ldr A.H. Orlebar, Flg Off D. D'Arcy Grieg and Flt Lt G.H. Stainforth. All three experienced engine cut-outs, but despite round-the-clock work faults with the Napiers could not be traced in time. Both Gloster VIs had to be withdrawn from the contest, yet ironically on the day following the contest (won by Great Britain with a Supermarine S6), Stainforth flew N249 over a measured course off Calshot and, in five passes, established a new World Air Speed Record of 336.3 mph. However, Sqn Ldr Orlebar exceeded this later in the Supermarine S6.

One Gloster VI appeared in London at the 1929 Lord Mayor's Show, and a year later was exhibited at the annual RAF Hendon Air Pageant. Afterwards, although officially attached to the High Speed Flight as trainers, the Gloster VIs rarely flew.

GANNET

Events which brought together a number of ultra-light aircraft from several manufacturers were the 1923–4 Motorized Glider competitions. These were sponsored by the *Daily Mail*, which early in 1923 offered a £1,000 prize, to which was added £500 by the Duke of Sutherland, for the pilot who made the longest flight on one gallon of petrol. The winning aircraft had to be powered by an engine of not more than 750 cc capacity, and in response a Light Aeroplane Competition was set up by the Royal Aero Club at Lympne aerodrome in Kent during October 1923.

Gloster's entry, the Gannet, emerged as one of the smallest aircraft ever built in Britain up to that time. With a length of 16½ ft and a span of 18 ft it weighed 283 lb. Of mainly wooden construction the Gannet's flat-sided fuselage was ply-covered with fabric over the top decking. A single-bay biplane, the upper centre-section contained a two-gallon fuel tank, while the wings folded back. The trailing-edge of the top centre-section folded up and forward to allow access to the cockpit. Wings were fabric-covered and ailerons fitted to all four. The engine was a 750 cc Carden 2-cylinder, air-cooled two-stroke, and the Gannet in GAC's finish of blue fuselage, with wings, tail unit and fuselage registration letters (G-EBHU) in ivory.

Test pilot Larry Carter had the Gannet airborne on 10 October, but the new and unproven Carden engine gave trouble and GAC was unable to compete seriously in the 1923 trials. During 1924 the Carden engine was replaced by a 7 hp Blackburne Tomtit which gave the

Gloster Gannet ultra-light (G-EBHU), with 7 hp Blackburne Tomtit engine and wings folded. GAC blue and ivory colours. (Courtesy of the Gloucester Citizen*)*

Gannet a top speed of 72 mph. This diminutive Gloster biplane entered the 1924 Lympne trials, but no details of its performance at those events are available as far as is known. Although not entered in futher competitions, Gloster's Gannet was maintained in airworthy condition but rarely flew. It appeared in the Aero Show at Olympia during July 1929.

GROUSE

The connection between Folland's Nighthawk, its derivatives and early GAC designs became manifest when a private venture built by the company emerged in 1923. Although classed as a two-seat training biplane, it was used by Folland to try out the in-flight behaviour of a new upper and lower wing combination. His idea was to bring together in one aircraft the benefits of both monoplane and biplane wings. Known as an HLB section, this set of wings comprised a single-bay layout with a high-lift upper mainplane and medium-lift lower mainplane. This resulted in good lift at take-off speed, while the lower wings, being at a smaller angle of incidence than the upper wing and of thin section, produced minimum drag, loss of lift at high speed and something like monoplane performance, all achieved without an increase in wingspan. With its relatively short fuselage and 230 hp Bentley rotary engine, this was a highly manoeuvrable biplane which GAC named the Grouse.

A modified version of Gloster's Sparrowhawk, the Grouse had its front cockpit faired over for initial trials, and with its HLB wing combination produced the most efficient lift/drag ratio experienced in any biplane up until that time. This design obviously possessed great potential as an interceptor, so much so that the Air Ministry requested a demonstration of the type at RAF Hendon. The Grouse (G-EAYN) arrived in GAC's blue and ivory colour scheme, and its display made quite an impression on the official observers. An Air Ministry contract followed for GAC to build three more of the type as Thick-Wing Nighthawks to be powered by a 350 hp Armstrong Siddeley Jaguar engine. The first machine (J6969) flew to the 1923 annual RAF Hendon Air Pageant, where it appeared in the New Types Park as the Gloster Grebe prototype.

During 1924 G-EAYN was returned to Sunningend for complete refurbishment, emerging early

Gloster Grouse II at Hucclecote, with Lynx engine, constant-chord ailerons and oleo landing gear. (GAC)

in 1925 as a two-seat trainer with a 185 hp Armstrong Siddeley Lynx radial engine. Designated Grouse II it incorporated several features of Grouse I and the Grebe, a number of components being identical on all three types. Indeed Folland used the original SE5a-type vertical tail surfaces of his Farnborough days in preference to the earlier Nighthawk/Mars-type fin and rudder.

The Grouse II featured external push-rods between the wings connecting both sets of ailerons, the Lynx engine was encased by an aluminium cowling provided with ports through which the seven cylinders protruded, and a streamlined propeller spinner was fitted. Two gravity-feed fuel tanks were located beneath the upper wing – one each side of the centre-section, landing gear was conventional with V-struts and cross-axle and the tailskid was steerable. Later updating included oleo front legs and extended rear stays for the landing gear, redesigned fuel tanks and constant-chord ailerons.

The Grouse proved extremely responsive to controls and very manoeuvrable, and when a replacement trainer for the Avro 504 was sought by the Air Ministry GAC entered the Grouse as a competitor, but it was unsuccessful and not accepted for RAF service. However, considerable interest was shown in the type by the Swedish Army Air Service, and when G-EAYN was shipped to Sweden the Army purchased it immediately. An order expected for eight more Grouse did not materialize, the prototype Grouse II as No. 62 remaining in Swedish Army service and operating with wheeled and ski landing gears.

Grebe

Following its debut at the 1923 RAF Hendon Air Pageant, Gloster Grebe prototype, J6969 (the renamed Folland Nighthawk Thick-Wing), flew to the A&AEE, Martlesham, for its official trials. Test and service pilots there reported the Grebe as being far in advance of any single-seat fighter developed at that time, with a top speed of 152 mph at sea level, a marked improvement over Sopwith Snipes then in RAF service. As a result, in addition to the three prototypes, an Air Ministry contract was awarded to GAC for twelve Grebe Mk II production aircraft (J7283–J7294) powered by a 400 hp Armstrong Siddeley Jaguar IV.

Of all-wooden construction with basically a box girder-type fuselage profiled by formers and stringers, the Grebe had no upper centre-section as port and starboard mainplanes were joined at the aircraft's centre-line supported by a cabane of inverted V-struts. Ailerons were connected by interplane push-rods, and production Grebe IIs had oleo-type landing gear, a steerable tailskid and redesigned fuel tanks.

The Grebe was Folland's first GAC fighter design built in quantity for the RAF, and joined Armstrong Whitworth Siskin IIIs in re-equipping the service's fighter squadrons in 1923. Grebes went initially to one flight of 111 Squadron at Duxford in October 1923, and during 1924 the type fully equipped 25 Squadron at Hawkinge replacing the unit's Sopwith Snipes.

A further four batches of Grebes were ordered for the RAF (J7357–J7402, J7406–J7417, J7519–J7538, J7568–J7603) plus a small contract for three machines (J7784–J7786), a total of 129 production Grebe IIs. They served with 19, 29, 32 and 56 Squadrons, and until 1928 Grebes were a regular feature at annual RAF Hendon air displays. In the 1925 display 25 Squadron gave an exceptionally well-disciplined programme of aerobatics and wing drill, during which the Grebes received some instructions by radio telephony from HM King George V who was at the show. In the 1931 air display three Grebes from the A&AEE gave a very polished performance of aerobatics when they created trails of coloured smoke. This was the last public appearance of Gloster Grebes in Britain.

One Grebe (J7519) was converted as a two-seat trainer, a small number of other Grebes being similarly altered as dual-control training aircraft. It was one of these two-seaters (J7520), with 39 painted on its rudder, which entered and won the 1929 King's Cup Air Race. It was sponsored for the event by the Member of Parliament for Cheltenham at the time, Sir Walter Preston, and flown by Flt Lt R.L.R. Atcherley with Flt Lt G.H. Stainforth as navigator. The average speed of the J7520 over the entire 1,160-mile course was 150.3 mph, a record at that time in the King's Cup Race.

A 'parasite' fighter experiment in 1926 featured a pair of Grebes; the concept was for two single-seat fighters to be suspended beneath the British airship R33, carried aloft and released in mid-air. These two Grebes (J7385 and J7400) were equipped with quick-release gear, extra bracing and carried no armament. The Grebes' Jaguar engines were started by a Bristol Gas Starter, housed within the airship and connected to both aircraft by flexible pipes. Special struts were fitted to the airship's bottom to steady the Grebes in transit, the aircraft itself being supported by a sling located above the centre of the upper wings. This sling pulled the Grebe up to the airship's supporting struts when attachment to the R33 was taking place. On 21 October 1926 both Grebes were successfully launched from R33, the first flown by Flg Off Mackenzie-Richards who dropped away from the airship at 10.17 a.m. The second Grebe was released at 11.40 a.m. over Cardington, both machines then making a normal landing (unlike US Navy fighters on similar experiments Grebes could not rehook to the airship in flight).

One Grebe, with a supercharged Jaguar engine fitted, achieved a top speed in level flight of 165 mph at 10,000 ft, and climbed to 20,000 ft in 16 minutes. Another Grebe, undergoing trials at the A&AEE, made a terminal velocity dive at 240 mph and, with only one bracing wire stretched, became the first British fighter to survive such a performance. Grebes were prone to wing flutter, however, and to counteract this V-struts were fitted between the wings outboard of the main interplane struts. In spite of this Grebes proved popular with their pilots, who appreciated the type's light, very responsive controls. Armament comprised two fuselage-mounted .303 in Vickers guns synchronized to fire through the propeller disc. RAF Grebes carried the ostentatious squadron markings of RAF fighters between the wars; the first to have theirs applied was 29 Squadron, with three large red crosses and continuous bars above and below carried on the

First production Gloster Grebe fighter (J7283) at Hucclecote in 1923. Coopers Hill, Brockworth, is in the background. (GAC)

fuselage sides and tops of upper wing surfaces. Grebes remained in first line RAF service until 1928, 25 Squadron being the last to exchange its Grebes for Siskin IIIAs at the end of that year.

Following New Zealand's decision to form a Permanent Air Force (NZPAF) it chose three Gloster Grebes as a nucleus. The first (ex-J7381), refurbished at Henlow, was shipped to New Zealand as NZ-501, making its initial flight there on 2 March 1928 from Wigram Air Base. Second and third New Zealand Grebes (J7394 and J7400) became NZ-502 and NZ-503 respectively. In two-seat form J7400 crashed during an aerobatic display on 8 August 1932, both crew members receiving serious injuries. When the NZPAF reorganized as the Royal New Zealand Air Force (RNZAF) Grebes NZ-501 and NZ-502 became A-5 and A-6 respectively with A Flight at Wigram. They remained in service for a number of years and, on 4 June 1938, were part of a simulated attack on a formation of RNZAF Vildebeest bombers. The following November both Grebes were withdrawn from flying duties to be relegated as ground instructional airframes, A-5 (INST.1) and A-6 (INST.2), the latter machine going to the Hobsonville Engineering School.

Grebe demonstrator, G-EBHA, first flew on 6 July 1923 with a 385 hp Jaguar IIIA and took part in the 1923 King's Cup Air Race as scratch machine. Pilot Larry Carter won a £100 prize for fastest circuit, but had to retire G-EBHA with a broken flying wire. This Grebe was rebuilt in 1926 and fitted with a 425 hp Bristol Jupiter IV engine. It featured modified landing gear, revised tail unit and upper ailerons, and became test-bed for the Gloster-Hele-Shaw-Beacham variable-pitch propeller.

GAMECOCK

The fighting cock bird is most significant to 43 Squadron of the RAF as it relates to a Gloster-built fighter biplane with which it began to equip in 1926 – the Gamecock. The squadron adopted the profile of a fighting cock bird as its crest to commemorate the unit's employment of Gloster Gamecocks in the mid-1920s. Ever since then 43 Squadron has been known as the 'Fighting Cocks', its markings being black and white chequers.

Folland designed the Gamecock to Air Ministry Specification 37/23, and in prototype form it appeared very similar to its Grebe predecessor. However, although retaining Folland's SE5 tail unit, the Gamecock was a completely new aircraft. The HLB wing had contributed much to the Grebe's speed and great manoeuvrability, but its heavy two-row Jaguar IV engine was the type's Achilles' heel. With the advent of Bristol's Jupiter IV engine Specification 37/23 actually required an updated Grebe powered by the lighter and less complex Bristol Jupiter engine.

Ordered as a Grebe II in August 1924, the prototype of the new design (J7497) was tested at Martlesham in February 1925 with a 398 hp Bristol Jupiter IV and an unbalanced SE5/Grebe rudder. The latter was changed for the more familiar horn-balanced type a few months later and two further prototypes followed, J7756 and J7757 with a Jupiter IV and 425 hp Jupiter VI respectively.

After extensive trials at Hucclecote and the A&AEE, during which no mention was apparently made regarding abrupt spinning tendencies or wing flutter (as had previously occurred on Grebes), an Air Ministry contract was awarded to GAC for thirty Gamecock I fighters in September 1925 (J7891–J7920) to be powered by Jupiter VI engines. Most of the first production batch of Gamecocks went to 23 Squadron, Henlow, in May 1926, this unit retaining its Gamecocks until July 1932, when they were exchanged for Bristol Bulldogs long after other Gamecock squadrons had been re-equipped. During the 1931 RAF Hendon Air Pageant, two 23 Squadron Gamecocks gave a very polished display of aerobatics piloted by Flt Lt M.M. Day and Plt Off Douglas R.S. Bader, who was to become the legless air ace of the Second World War.

A further 42 Gamecocks were ordered in July 1926 (J8033–J8047 and J8069–J8095), and the following November a contract was signed for another eighteen machines (J8405–J8422).

Gloster Gamecock (J8033), first of the second production batch, on Hucclecote airfield in 1926. In the background a hedge and some houses line what is now the busy A417 trunk road. (GAC via BAe plc, Manchester)

Aircraft from these batches went mainly to 32 and 43 Squadrons, and the type also served for a short time with two night interceptor units, 3 and 17 Squadrons at Upavon, where they flew as night fighters until replaced by Bristol Bulldogs in May 1929.

Gamecocks were the RAF's last all-wooden biplane fighters, with ash longerons and spruce, ash or plywood struts. Internal bracing was by steel tie-rods, and the engine attached to the steel bulkhead by nine large bolts. The pilot sat in an open cockpit, his armament comprising two .303 in Vickers machine-guns synchronized to fire through the propeller disc.

Gamecocks inherited the Grebe's wing flutter problem and lacked remedial stability in a spin. In their first nineteen months of RAF service twenty-two Gamecocks crashed with the deaths of eight pilots. After many tests at Hucclecote and Farnborough, interplane V-struts were installed outboard of the N-struts as on the later Grebes, and this helped cure the problem. Even so Gamecocks were very prone to torque and right-hand spins were banned. Some pilots did carry out prolonged left-hand spins, and GAC test pilot, Capt Howard Saint, demonstrating the effect of the new V-struts, dived a Gamecock steeply at 275 mph and pulled out without any problem.

A number of Gamecocks were used experimentally; J8047 had a lengthened fuselage, revised rudder, modified ailerons, wide-track landing gear, Jupiter VII engine and trials with a Hele-Shaw-Beacham variable-pitch propeller. Unofficially known as Gamecock III this aircraft was sold as scrap in April 1934, but was acquired by J.W. Tomkins, when it was refurbished as G-ADIN and flew until September 1936. Gloster's Gamecock demonstrator (G-EBNT) was standard, but fitted with a streamlined spinner and modified engine cowling. Gamecock J8075 flight-tested the Bristol Mercury IIA engine, while J8804, designated Mk II, had its rudder and ailerons modified. Although an improvement over previous Gamecocks, no production Mk IIs were ordered.

However, Gamecock II interested the Finnish Government, and two pattern aircraft were purchased from GAC. Within six months licence-built Gamecocks were being built at Helsinki provided with interchangeable wheel and ski landing gears. In Finland Gamecocks were known as the Kukko, and one Finnish Air Force unit flew its Kukkos from 1929 until 1935. Indeed one machine (GA-46) remained in service until 1944! Finnish Kukkos were powered by Gnome-Rhone-built Bristol Jupiter VI engines.

Gamecock Derivatives

As GAC design staff were heavily involved with the Gamecock and Gloster II racing seaplane in 1924, they were hard pressed to concentrate on a £24,000 Air Ministry contract received that May to build three experimental single-seat fighters powered by Napier Lion liquid-cooled engines. Thus a year elapsed before the first machine (J7501) appeared as the Gorcock with a 450 hp Lion IV installation. The first two Gorcocks were to have steel fuselages and wooden wings, while the third was to be an all-steel affair.

When the first machine appeared its Gamecock ancestry was apparent, but of greater aesthetic appeal. The Lion engine was beautifully installed, its three banks of 'W' cylinders faired smoothly into the fuselage and complemented by a streamlined propeller spinner. Upper wing fuel tanks were retained, but with minimum protuberance, and although Folland returned to his SE5/Grebe empennage initially, it was soon replaced with the Gamecock horn-balanced-type rudder. The Lion drove a two-blade fixed-pitch wooden propeller, a Gamecock style landing gear was used and armament comprised two .303 in Vickers machine-guns synchronized to fire through the propeller disc from fuselage troughs. Four 20 lb bombs could be carried on underwing racks.

The second Gorcock (J7502) was similar to J7501, but had a 525 hp Napier Lion VIII direct-drive engine. Numerous problems with the Lion engines in relation to required performance figures meant neither Gorcock could be delivered to the Air Ministry until 1927. In June that year the third, all-metal, Gorcock (J7503) was completed and reverted to a 450 hp geared Lion IV. It joined the other two Gorcocks at Farnborough, where all three machines were employed for a number of years on research and development flying.

The Air Ministry also placed a £15,000 order with GAC for three experimental single-seat high-altitude fighters to test supercharged engines. Named Guan (a large noisy American tree-dwelling game bird), this design was based on the Gorcock and Folland used his HLB wooden wings with an extra 3 ft 4 in span added. The Guan was intended to combine the Gorcock's top speed (30 mph faster than contemporary RAF fighters) with a higher service ceiling, which it accomplished easily.

The first Guan (J7722) had a 450 hp supercharged Napier Lion IV with an exhaust-driven

Second Gloster Gorcock (J7502) at Hucclecote in 1927, with direct-drive Napier Lion VIII engine, modified cowling, flush exhaust pipes, restyled top decking aft of cockpit and Gamecock type rudder. (GAC)

turbocharger located externally below the propeller shaft, an arrangement that produced excessive pipework and protuberance around the nose section. This Guan, completed in June 1926, was delivered to Farnborough to be followed by the second machine (J7723) early in 1927. This differed in having a 525 hp direct-drive Napier Lion VI for which an exhaust-driven turbo-supercharger was mounted atop the cowling above the propeller shaft. These superchargers enabled the Guans to maintain maximum power up to 15,000 ft, where the top speed was 175 mph and service ceiling 31,000 ft. However, many problems arose with the superchargers and, despite efforts by Napier's to modify and improve the system, abandonment of the programme was inevitable and the third Guan, which was to have had a supercharged Napier Lioness engine, was not built.

In 1926 Air Ministry Specification F9/26 called for a metal successor to the wooden Gamecock. GAC was awarded a contract to construct an all-metal version of the Gamecock as a high-altitude day and night fighter. It was named Goldfinch.

Similar in profile to Gamecock II, the original Goldfinch (J7490) featured a mixed wood and metal fuselage, the wings and tail unit being all-metal. A 450 hp Bristol Jupiter VIIF radial engine was fitted, and the pilot sat in an open cockpit with a good all-round view. Folland revised the fuselage by increasing its length and employing an all-metal structure, and the tail unit was akin to that of the experimental Gamecock III, J8047.

In its modified form the Goldfinch flew to Martlesham in December 1927, where it achieved 172 mph at 10,000 ft, climbed to 20,000 ft in sixteen minutes and recorded a service ceiling of 27,000 ft. But Gloster's hopes for a production order were dashed when the Goldfinch failed Air Ministry requirements in relation to fuel capacity and military load-carrying capabilities. This competition was won by the Bristol Bulldog.

Gloster's international reputation as a producer of high performance military aircraft was enhanced by another Gamecock derivative, the Gambet. This was inadvertently the result of an Imperial Japanese Navy request to Japanese aircraft manufacturers Aichi, Mitsubishi and Nakajima for a new fighter to replace the Japanese Navy's ageing Gloster Sparrowhawks.

Nakajima approached GAC with the intention of acquiring manufacturing rights for the

Powered by a Jupiter VIIF engine, this was Gloster's one-off Goldfinch (J7940) of 1926. (Rolls-Royce plc)

Gamecock, but Folland was already working on a private venture shipboard fighter variant of the Gamecock. He hoped it would interest the British Air Ministry, while at the same time provide a successor for the Japanese Sparrowhawks. The Gambet roused little interest at the Air Ministry; Japan's Navy jumped at it. Nakajima obtained their manufacturing rights and a pattern aircraft was shipped to Japan. Nakajima's chief designer, Takeo Yoshida, modified the Gambet to suit Japanese production methods, replacing the Bristol Jupiter VI engine with a 520 hp Nakajima licence-built Jupiter. GAC had already installed night flying equipment.

The Japanese Gambets became Navy Type 3 Carrier Fighter Model 1s, or A1N1s, and later modifications resulted in the A1N2 with a 520 hp Nakajima Kotobuki 2 engine, metal propeller and sliding cockpit canopy. These A1N1s and A1N2s were reported as the best carrier fighters then employed by the Japanese Navy and, during the conflict with China which started in 1932, they proved superior to any enemy aircraft types.

DH9As AND GORAL

Derived from the DH9, a two-seat bomber that had proved a disappointment because of its Siddeley Puma engine, the DH9A started as a converted DH9 fitted with a 375 hp Rolls-Royce Eagle VIII engine. Eventual standard powerplant for DH9As was, however, the American 400 hp Liberty 12, chosen for the DH9 because of a huge demand for Rolls-Royce Eagle VIIIs. Liberty-powered DH9As became an outstanding strategic bomber used by the 1918 Independent Air Force, carrying out several daylight attacks on German targets. After the war they flew with 47 and 221 Squadrons in Russia during the 1919–20 fight against the Bolsheviks, and for well over a decade DH9As were a mainstay of RAF units in Iraq, over the North West Frontier of India, at home with the Auxiliary Air Force and as trainers in a number of RAF flying training schools.

Affectionately known as 'Nine-Ack' the DH9A was an all-wooden two-bay biplane with a two-man crew sat in tandem open cockpits. Armament comprised a .303 in Vickers machine-gun synchronized to fire through the propeller, and a .303 in Lewis gun on a Scarff gun-ring in the rear cockpit. Up to 660 lb of bombs could be carried on underwing racks.

Example of a de Havilland DH9A, a type refurbished and newly built by Gloster's in the 1920s. (Author's collection)

Based on DH9A components, this was Gloster's one-off Goral (J8673), which had a Jupiter VIA engine. (Rolls-Royce plc)

A share of an Air Ministry contract for 120 refurbished DH9As (J7008–J7127) went to GAC at Sunningend, where fifteen were reconditioned and a further ten rebuilt from stored airframes in the serial block J7327–J7356. Ten new machines were also ordered from GAC (J7249–J7258). It is known that one Gloster-rebuilt DH9A (J7347) was with 27 Squadron in India. An offshoot of the DH9A emerged from GAC in response to Air Ministry Specification 26/27 calling for a general purpose replacement for DH9As and Bristol F2Bs. Stipulations included a capability to undertake special tasks at home and abroad, improved load-carrying and performance, and incorporation of as many DH9A components as possible to reduce a surfeit of those in Ministry-held stock. An all-metal fuselage would be more durable in hot climates and a Napier Lion engine preferred (also large numbers on Ministry charge) but not compulsory.

Gloster's answer was the Goral, designed by Capt S.J. Waters under Folland's jurisdiction. A number of DH9A parts were used, the main items being a complete set of wings and fittings. The empennage differed in having a distinct angular profile. The rest of the airframe was metal with a fabric covering, the fuselage being in three sections for ease of transport and erection overseas. GAC did not use a Napier Lion engine, but installed a 425 hp Bristol Jupiter VIA direct-drive radial, with the intention that if the Goral was accepted for service the Jupiter VIA would be replaced by a geared 580 hp Jupiter VIII.

In the event Gloster's Goral (J8673) competed against six prototypes from other manufacturers, Westland winning with its famous Wapiti. GAC did not lose out altogether, however, for the Company's expertise in metal airframes won it a contract to provide over 500 sets of wings for Wapiti Mk IIs.

Goring

An attempt to break away from Gloster's racing and fighting aircraft designs was made by Capt Waters and Folland when they designed a two-seat day-bomber/torpedo-bomber biplane as a private venture. Built at Sunningend, this aircraft was later entered in a competition to Air Ministry Specifications 23/25 and 24/25 seeking replacement for the Hawker Horsley.

Numbered J8674 and named Goring, Gloster's new bomber made its initial flight from Hucclecote in March 1927, and although of orthodox single-bay biplane configuration, its clean lines revealed some of Folland's earlier concepts used in designing his racing seaplanes. Anhedral applied to the lower wing roots, in similar fashion to the Gloster IV racer, gave an inverted gull appearance, improved aerodynamic efficiency and allowed a short robust landing gear to be fitted. This alternated as a cross-axle, or divided type for torpedo-carrying purposes, and was convertible to seaplane configuration using twin floats designed and built by Gloster.

A Joukowski aerofoil section with very high lift properties was employed in the Goring's mainplanes, their relatively thick section permitting two 75-gallon fuel tanks to be contained within the upper wings. The Goring airframe was composed mainly of wood and steel fittings, but a production version would have an all-metal framework with a fabric covering. The pilot's seat and rudder pedals were adjustable, and a .303 in Vickers gun in the port side was synchronized to fire through the propeller. To provide maximum rear defence the observer/gunner's seat was on rollers inserted into slots fitted to the fuselage sides. To allow the gunner space to operate the .303 in Lewis gun mounted on its Scarff ring, or gain access to a bomb sight or second machine-gun in the floor, this seat could be quickly stowed.

The Goring was powered by a 425 hp Bristol Jupiter VI direct-drive radial. Although the intention had been to fit a 470 hp Jupiter VII with gear-driven supercharger, on test this had proved unreliable as had the Bristol Orion engine with an exhaust-driven supercharger. By the time it appeared at the 1927 annual RAF Hendon Air Show, the Goring had been re-engined with a 460 hp geared Jupiter VIII. In 1928 the Goring gained quite a reputation for its range and load-carrying capabilities at the A&AEE, where it competed with the Handley Page Hare, Hawker Harrier and Westland Witch. In the event none of the four contenders met Air Ministry requirements due to setbacks with Bristol's Jupiter VIII and X engines and superchargers.

Returning to GAC in 1930 the Goring underwent a conversion to seaplane configuration and was fitted with an enlarged rudder. Gloster's test pilot Rex Stocken carried out extensive trials with the Goring seaplane at RAF Calshot, some of these including full-load conditions with

Ultimate version of Gloster Goring J8674, with redesigned empennage, uprated Jupiter XF engine and new spinner. (GAC)

underwing bomb racks. These carried either four 112 lb, two 230 lb or 250 lb, or sixteen 20 lb bombs. After its seaplane trials J8674 was returned to GAC for reconversion to landplane form. A redesigned fin and rudder was introduced and the Bristol Jupiter VIII replaced by a 575 hp Jupiter XF. Later taken on Air Ministry charge, the Goring was moved to Filton and became a flying test-bed for Bristol's 745 hp Mercury VIIA, 570 hp Pegasus II and 670 hp Perseus sleeve-valve air-cooled radials.

THE BIG MOVE AND PROTOTYPES

As the use of metal in aircraft construction increased, GAC's Hugh Burroughes was convinced all-metal airframes held the future key to successful aircraft design. Some members of the Board were not so enthusiastic, however, especially those from H.H. Martyn, and after GAC chairman A.W. Martyn resigned in September 1927 his place was taken by David Longden, a keen supporter of the metal airframe concept. Hugh Burroughes also persuaded fellow Board members it would be a wise move for him to buy half the shares in the Steel Wing Company, which had been formed in 1919 for research into the use of steel in aircraft. During 1928 Burroughes then sold his Steel Wing shares to GAC, and the Steel Wing Co. became part of Gloster Aircraft.

With this expansion the aforementioned move to Hucclecote from Cheltenham was a necessity as new hangar and office accommodation would be required. Until 1927 GAC had been content to rent two hangars, but further buildings had to be rented and in 1928 the company decided to purchase the complete Hucclecote site of 200 acres, including all hangars and offices.

The acquisition of the Steel Wing Company proved very beneficial to GAC, for it was entrusted with several important contracts, securing employment for employees from 1927 until 1934 at a time when no production aircraft was forthcoming from the Gloster company. The first contract was for seventy-four metal-structured Armstrong Whitworth Siskin IIIA single-seat fighters for the

This Gloster-built Armstrong Whitworth Siskin IIIA (J8959) is from 43 Squadron, the 'Fighting Cocks' (black & white chequered insignia), and flown by the Commanding Officer Sqn Ldr Lowe. (Whitworth-Gloster Aircraft Ltd)

RAF, built between 1927 and 1929. A follow-up contract was for a large modification and repair programme on over 100 Siskins, and from 1929 until 1932 a total of 525 sets of metal wings for Westland Wapiti IIA biplanes were constructed under contract in No. 2 Hangar at Hucclecote.

A one-off contract in 1929 was for the construction at Hucclecote of an Italian variable-camber wing to be fitted experimentally to an Italian Breda 15 high-wing monoplane. This wing's trailing edge was built in three sections, with ailerons at the tips, inboard of them a cable-operated variable-camber portion, and at the roots a movable section with an up and down limited motion. The first flight from Brockworth was on 8 September 1933 by an Italian pilot, and GAC test pilot Rex Stocken flew it during November. But on 1 December, while being flown by GAC chief test pilot Howard Saint, the Breda (G-ABCC) developed wing flutter and crashed on Churchdown Hill. Saint was unhurt, but the aircraft was a write-off.

Under another contract GAC completed de Havilland's DH72, a gigantic tri-motor night bomber, which its parent company had commenced building in 1928. Serial number J9184, this large biplane was powered by three 595 hp Bristol Jupiter XFS radials, and its wings spanned 95 ft. There were four main landing wheels, twin fins and rudders, with defensive gun positions located in the nose and tail. Weighing 21,460 lb, J9184 made its first flight from Brockworth in 1931. Earlier that year another de Havilland machine, the DH77 low-wing monoplane Interceptor (J9771), had been sent to GAC for flight trials and a 100-hour development programme on its Napier Rapier 1 'H'-type air-cooled engine. This machine remained at Hucclecote until December 1932, when it transferred to the RAE, Farnborough.

Meanwhile H.J. Steiger had designed the monospar method of aircraft construction and formed the Monospar Wing Co. In 1929 he signed a contract with GAC to construct his design for a small twin-engined three-seat cabin monoplane powered by two Salmson engines. Known as the Gloster-Monospar SS1, this aircraft was flying by October 1930 and passed a number of very demanding tests. It was the progenitor of a successful line of Monospar monoplanes built by General Aircraft, who had taken over development and patent rights of Steiger's system. To prove this type of construction was suitable for large as well as small aircraft, GAC built a one-piece,

Powered by two Salmson engines, this Monospar ST3 (G-AARP), standing outside the main hangar at Hucclecote, was built by Gloster in 1930. (GAC)

63 ft-span wing to Air Ministry order for a Fokker FVIIb-3m tri-motor. Despite its size this wing in its completed form amounted to only 10 per cent of the Fokker FVIIb-3m's gross weight.

Gnatsnapper

Following success with the Gambet for Japan's Imperial Navy, Folland designed a single-seat carrier-borne naval fighter in 1926 which responded to Specification N21/26. The new design had to be capable of the top performance demanded by naval aircraft encountering new concepts in naval shipboard-operating methods. An all-metal airframe was essential, and in a rider it was stated that preference would be given to use of a 450 hp Bristol Mercury IIA engine (then new and untried).

Work commenced on the new aircraft at Hucclecote in June 1927. The fuselage was of tubular steel with light alloy formers, metal panelling covered the front portion and fabric the rear. For ease of maintenance the engine was on hinged mountings which allowed it to be swung either side. Wings of single-bay equal span biplane layout, based on GAC's lattice-type steel spar system, had ailerons fitted to the upper mainplanes only, which featured prominent dihedral. The empennage was conventional, but large horn balances were incorporated into the elevators and rudder. An orthodox V-strut and cross-axle-type landing gear was fitted with brakes on the main wheels. A tailskid was located below the rear fuselage.

Few problems arose with construction of what was named the Gnatsnapper (N227), but considerable delay occurred during the development of Bristol's Mercury IIA engine. On arrival at GAC the first proved overweight and unreliable during initial tests and the new fighter was fitted with a 450 hp Jupiter VII for its flight trials. Its first flight from Brockworth was in February 1928, but by May 1929, after six successive Mercury IIAs had been installed in N227, there was little improvement.

Too late for entry in the shipboard fighter competition, the Gnatsnapper was refitted with a

Here Gloster Gnatsnapper II (N227) is shown in the 1930 updated form, with revised empennage and Townend ring cowling for its Jaguar VIII engine. (BAe plc, Manchester)

Jupiter VII engine and sent to the A&AEE for further trials. In this form the fighter proved very satisfactory, and test pilots at Martlesham spoke highly of its excellent manoeuvrability, diving abilities and top speed of 165 mph at 10,000 ft. A modification programme carried out on N227 included manually operated double-camber wings, revised rudder profile with rounded top and greater area, a more robust landing gear to withstand the rigours of carrier-deck flying and a further Mercury IIA engine installation. However, among other manufacturers involved in the N21/26 competition, GAC was requested by the Air Ministry to revise its Gnatsnapper by fitting a 540 hp Armstrong Siddeley Jaguar VIII supercharged radial. To accept this engine the structure had to be altered, and the guns moved from the fuselage sides to a position on top ahead of the cockpit, the twin Vickers guns now firing from troughs in the front upper decking.

In January 1929 work started on a second Gnatsnapper (N254), but this aircraft was not ready until March 1930. A Mercury IIA engine was fitted but still caused problems, so much so that Gloster's finally rejected it for a Bristol Jupiter VII fitted with an exhaust collector ring. By April 1930 the first machine (N227) was back at Hucclecote in order to fit plain ailerons and a new oval fin. The engine was a Jaguar VIII, and the aircraft redesignated Gnatsnapper II. In December 1930 it arrived at Martlesham for trials, but was badly damaged in a landing accident there and returned to Gloster's. Repairs and further improvements were carried out, a Townend ring cowling fitted around the Jaguar engine, and the aircraft used on armament trials until summer 1931, when it returned again to GAC for major modifications. These included two-bay wings in place of the single-bay type and substitution of the Jaguar engine by a 525 hp Rolls-Royce Kestrel IIS steam-cooled, geared and supercharged, V-type. Steam condensers were incorporated along the leading edges of the upper and lower wings.

Redesignated Gnatsnapper III, N227 now featured a distinct streamlined profile, but any service machine engaged in combat with such a cooling system would be extremely vulnerable. Thus this idea was abandoned, further trials at the A&AEE proving negative. The Gnatsnapper eventually passed to Rolls-Royce at Hucknall and became a flying test-bed for the 600 hp Rolls-Royce Goshawk engine. Later N227 was used by Rolls-Royce as a company 'hack', finally going to the scrap heap in 1934.

GLOSTER AS31 SURVEY

With Great Britain maintaining a vast empire during the 1920s, it was considered her Dominions and Colonies could be further expanded and developed by aircraft flying over remote areas and mapping out the regions. Much of this hazardous work was over extremely hostile terrain in Asia, Africa, South America and the Arctic, and was carried out by single-engined DH9s, many of which were war-surplus machines. To force-land in a remote area with such an aircraft with potential engine trouble could be fatal.

A new aircraft, purpose-built for surveying, was ordered with the stipulation that power must be provided by at least two engines. To facilitate emergency ground transportation the aircraft should be sectional, and emphasis was placed on the photographer, for whom partially enclosed cabin accommodation was essential. This position must have a good field of view, facilitate instrumental observations and contain a means of communication with the pilot.

Design and construction of the new aircraft was undertaken firstly by the de Havilland Aircraft Co. as the DH67, work starting on the machine in 1927. Owing to preoccupation with the DH66 airliner and famous Moth series, however, de Havilland transferred work on the survey aircraft to GAC at Hucclecote under a contract of November 1928. Initially a small version of the de Havilland DH66 airliner, the DH67 had twin engines and single vertical tail surfaces, whereas three engines and triple fins and rudders were a feature of the DH66. At Hucclecote Folland modified the design to accommodate eight passengers, or it could be used as a flying ambulance, cargo plane, or bomber armed with three machine-guns. GAC called this much-updated layout the Gloster AS31 Survey.

Of all-metal fabric-covered construction, with a square-section fuselage built in three parts, the AS31 could easily be dismantled into portions by removing a few bolts and pins, and reassembly could be undertaken by semi-skilled labour. Wing construction was of the Gloster lattice spar-type, and ailerons were fitted to the lower wings only. The tailplane was fully adjustable in flight, and Folland designed a wide-track divided-type landing gear using a triangle and 'V' layout; the rear fuselage was supported by a tailskid. Dual controls were fitted and, to ensure more positive control as well as extended life to the system, push-rods superseded the normal cables. An increase in power came from two 525 hp Bristol Jupiter XI geared radials, although Folland did allow for alternative engines to be installed (Armstrong Siddeley Jaguar Major or VI, American Pratt & Whitney Hornets, Wright Cyclones or French Lorraine-Dietrich 14ACs). Camera equipment included a Williamson Eagle on a special mounting which enabled it to be lowered through a purpose-built aperture in the fuselage bottom. This was an asset for photography over a wide radius, and kept the camera free from any obstruction on the aircraft.

Gloster's produced two AS31s, the first (G-AADO) making its initial flight in June 1929 piloted by Capt Saint. That first venture into the air was a little premature as the machine was only supposed to be on taxying trials. Folland was in the cockpit with Saint who, having taxied to the end of Brockworth airfield, turned the aircraft and opened the throttle to avoid some ruts. Suddenly G-AADO was airborne and Saint took her up steadily, circuited the airfield and landed. A second AS31 (K2602), built for the Air Ministry, was an exhibit at the International Aero Show, Olympia, in July 1929. The following November it went to Farnborough, where it flew on wireless and telegraphy experiments, remaining there until struck off charge in March 1937.

Meanwhile G-AADO was successfully demonstrated to Aircraft Operating Co. directors at Heston on 25 January 1930. On 20 March G-AADO left on a 7,000-mile flight to Cape Town, in conjunction with a Zambezi Basin Survey Expedition, landing at Wyabird airfield on 11 April after averaging 128 mph on its flight from the UK. Later in 1930, at Bulawayo, G-AADO participated in the Cochran-Patrick survey and, within a year, had surveyed

The Gloster AS31 Survey biplane (G-AADO) on Hucclecote airfield in 1929. Note the mudguards over the wheels. (GAC)

63,000 square miles of Northern Rhodesia. It had clocked some 500 hours flying time, and apart from tyres and tailskid shoes no parts needed replacing. Aircraft Operating Co. director Alan Butler later wrote to Gloster praising the aircraft for its reliability and performance. Eventually this AS31 was sold to the South African Air Force, when it became No. 250. Based at Zwartkop aerodrome in Valhalla, it was used on aerial photography duties until 1942!

TC33 – Gloster's Jumbo

Air Ministry Specification 16/28 called for an aircraft capable of carrying thirty fully armed troops, an equivalent weight in cargo, or 6,000 lb of bombs. An allowance for 2,000 lb of fuel and oil over a distance of 1,200 miles non-stop was to be made.

Gloster responded with the largest (not heaviest – that would be the Javelin) aircraft to ever be rolled out of the Hucclecote works. Sometimes named Goshawk, but more generally known as the TC33, Folland's design emerged as a bomber-transport with serial number J9832. It was unorthodox in having a greater lower wing-span than upper, 95 ft 1 in and 93 ft 11 in respectively, while the tailplane was a sesquiplane layout with an upper tailplane supported on struts above the extreme rear of the fuselage. Twin fins and rudders were fitted.

An all-metal affair the TC33 had a long oval fuselage with a pencil-like profile. Pilot, co-pilot and navigator sat in an enclosed cockpit, but the hapless front gunner was perched in the aircraft's nose in an open cockpit armed with a .303 in Lewis gun on a flexible mounting. Access to his position was via a small door located in front of and beneath the co-pilot. The main cabin was nearly 28 ft long, soundproofed, and carried thirty soldiers plus their equipment, or up to twelve stretcher cases if employed as an air ambulance. Used as a freighter it could convey more than 2 tons of cargo, a large drop-down door providing access for loading and unloading bulky or heavy items, which could be lifted into the fuselage by means of a mechanical hoist attached to a fuselage beam. Smaller freight was loaded by crane through a door in the cabin roof. A gangway ran through the rear fuselage to the tail gunner's position, an open cockpit located at the extreme rear of the fuselage with a .303 in Lewis machine-gun on a flexible mounting.

Massive fabric-covered mainplanes were fitted, unstaggered and of single-bay biplane layout, with construction based on Gloster's metal lattice spar and rib principle. Lower wings were in a mid-wing position in relation to the fuselage, and in order to avoid interior obstruction the centre-section possessed an acute anhedral angle which raised the inboard wing spars clear of the aircraft's main cabin. This gave the TC33's lower wing a distinct inverted gull look. The landing gear comprised a pair of hefty 5 ft diameter wheels and heavy duty tyres, carried on sturdy triangular structures with shock absorbers incorporated into the vertical members. Each main wheel was equipped with compressed air-operated brakes, and the wheel track was no less than 22 ft 6 in. The engines were four 580 hp Rolls-Royce Kestrel steam-cooled Kestrels strut-mounted in tandem pairs, tractors (Kestrel IIIS), pushers (Kestrel IIS). Atop each tractor unit a steam condenser provided evaporative cooling for the tandem pair. This system required 6 gallons of water against the 14 gallons needed for the average liquid-cooled aircraft engine, a weight-saving factor of 250 lb.

Work on the TC33 commenced at Hucclecote late in 1930. By January 1932 the big machine was ready, but the upper mainplanes were too high to pass through the doors of Hucclecote's No. 2 hangar. It was necessary to dig trenches for accommodation of the huge wheels before J9832 could be winched out! The initial flight took place on 23 February 1932 piloted by Howard Saint, and after modifications to rectify vibration problems had been carried out, official trials followed at the A&AEE. The next venue for J9832 was the annual RAF Hendon Air Pageant, where it was No. 6 in the New Types Park. A twelve-minute demonstration flight

The largest aircraft built by GAC, the TC33 bomber-transport (J9832) at Hucclecote in 1931. Note the steam condensers above each tandem pair of steam-cooled Kestrel engines. (Rolls-Royce plc)

was performed in front of the public, and two days later the TC33 gave a repeat performance at the SBAC, also at Hendon. As a bomber, J9832 carried its bombs on detachable bomb racks under the fuselage, a normal load consisting of twelve 250 lb and four 20 lb bombs, a total of 3,080 lb, which allowed a realistic fuel capacity to be taken into account.

In October 1933 J9832 was back at Martlesham, where 141.5 mph was recorded in level flight. Some reports were favourable and the standard of crew comfort found admirable (it is debatable if this applied to the front and tail gunner). But take-off performance fully loaded was poor, some take-offs during trials having to wait for a suitable wind! The massive landing gear was also criticized owing to its inconsistent behaviour between take-off and landing loads. In consequence it was decided the TC33 would be unsuitable for operations from RAF airfields, and J9832 remained the only example of GAC's jumbo built.

SS18/19 AND GAUNTLET

As the last open cockpit biplane fighter in RAF service, Gloster's Gauntlet owed its pedigree to Folland's previous line of biplane interceptors. Its progenitor was the SS18 built to Air Ministry Specification F20/27, which had called for a single-seat interceptor of primarily metal construction, with main emphasis on speed, rate of climb, manoeuvrability and good pilot view.

Folland's SS18 design was an all-metal two-bay biplane, with metal panelling between the cockpit and nose, and a fabric-covered rear fuselage and wings. Steel tubing featured throughout the fuselage, with high tensile steel main spars and steel ribs forming the mainplanes. The engine was a 450 hp Bristol Mercury IIA uncowled and, although not producing the 500 hp envisaged when designed, this powerplant, combined with the aircraft's streamlining, pushed the SS18 to a top speed of 183 mph at 10,000 ft. With serial number J9125 the prototype made its first flight in January 1929 piloted by Howard Saint.

Meanwhile Bristol's successful Bulldog fighter was entering RAF squadron service with a Bristol Jupiter VIIF engine. This prompted Folland to fit a Jupiter engine in the SS18, which then became the SS18A. It flew for the remainder of 1929 with the Jupiter, but in 1930 an Armstrong

With uncowled Mercury IIA engine, the Gloster SS18 stands outside GAC's Hucclecote works in 1929. Airframe J9125 was used throughout from this model to the ultimate SS19B Gauntlet prototype. (Rolls-Royce plc)

Siddeley Panther III was fitted encased in a Townend ring cowling. Redesignated SS18B the fighter could now achieve 205 mph at 10,000 ft, but the heavy Panther engine had an adverse effect on the climb rate, so the Jupiter VII was refitted, Townend ring cowling retained and J9125 redesignated SS19. For its official A&AEE trials some heavy armament had been incorporated into the design to comply with an Air Ministry demand for a fast multi-gun interceptor capable of combating any potential bomber threats to the UK. This armament comprised two .303 in Vickers machine-guns, firing from troughs in the forward fuselage sides synchronized to fire through the propeller disc, and four Lewis guns, two beneath the upper and lower wings respectively. Provision was also made to carry four 20 lb bombs under the lower wings.

Gross weight was now 3,500 lb, but with a Jupiter VIIFS fitted top speed was 188 mph at 10,000 ft. A&AEE report on the SS19 trials was satisfactory, but Air Ministry scepticism of the six-gun armament resulted in deletion of the four wing-mounted Lewis guns and installation of night flying equipment. This interchange obviated any appreciable difference in the weight factor and Folland added other refinements, including streamline main wheel spats, a similar small spat for the tailwheel which replaced a skid and increased fin area for better lateral control. Designated SS19A GAC's fighter now had a top speed of 204 mph at 10,000 ft, which was 30 mph faster than the Bristol Bulldog and within the speed range of Hawker's in-line-engined Fury biplane fighter. This was the SS19A's ultimate performance Jupiter-powered, and in the autumn of 1932, J9125 was fitted with a 536 hp Bristol Mercury VIS radial. It now became the SS19B, and was recognized as a true prototype for the Gloster Gauntlet, the name bestowed later in 1933.

A&AEE trials in the spring of 1933 proved quite successful, and that summer a 570 hp Mercury engine was installed surrounded by a Boulton & Paul ring cowling, which produced a top speed of 212.5 mph at 20,000 ft. In September 1933 a preliminary issue of Specification 24/33 was issued to GAC, calling for twenty-four production aircraft which were required for RAF service by March 1935. Confirmation of this order came in February 1934, by which time J9125 had been further updated with improvements, including an uprated 640 hp Mercury VIS2 with a narrow-chord cowling and prominent exhaust collector ring, a RAE Mk IIA air starter superseding the Hucks type, two Vickers Mk V machine-guns, Dowty oleo-type landing gear in place of the original Vickers, and deletion of the wheel spats, which clogged when operating from soft ground (some early production Gauntlets were still being produced with wheel spats months later).

The first production Gauntlet (K4081) made its initial flight from Brockworth on 17 December 1934 piloted by P.E.G. 'Gerry' Sayer, who had replaced Howard Saint as Gloster's chief test pilot. The performance figures showed further improvement, Gauntlet Mk I having a top speed of 230 mph at 15,800 ft. On 18 February 1935 Gauntlet I, K4086, was taken on charge by 19 Squadron, Duxford. After three months with the squadron on familiarization trials, K4086 was followed by other early production machines, which were delivered to 19 Squadron between May and June 1935 (this unit would not replace its Gauntlets with Spitfires until January 1939).

Meanwhile, as GAC developed the SS18/19, they faced serious financial problems. A programme of non-aviation activities was undertaken from around 1930 until 1934, involving the use of some hangars for off-season parking of motor coaches from Black & White Motorways of Cheltenham, pig farming, mushroom growing or as badminton and indoor tennis courts. Actual work entailed the production of such diverse items as milk churns, motor car bodies, roll-up shop fronts, gas-fired fish fryers and even a motorized barrow fitted with crawler tracks designed to carry bombs or stores over muddy surfaces.

This lean period came to an end in 1934, when GAC was taken over by the expanding Hawker Aircraft Co. Gloster lost its independence, but Hawker's full order books and an assurance that GAC employees could look forward to a long period of full employment on aircraft production induced Gloster's Board to accept Hawker's proposals (Hawker aircraft types built by GAC are covered separately). With Gauntlet Is already entering RAF service, an order for a further 104 of the type was issued in April 1935, but as Hawker Aircraft now had control of the company it was decided to adopt a Hawker construction system for the new Gauntlets. This chiefly involved revised methods of producing wing main spars and rear fuselages, and resulted in the Gauntlet Mk II, of which another 100 were ordered by the Air Ministry in September 1935.

The first Gauntlet IIs went to 56 and 111 Fighter Squadrons at North Weald and Northolt respectively, further deliveries being made to 17, 32, 46, 54, 65, 66, 74, 79, 151 and 213 Squadrons. One unit (80 Squadron) flew Gauntlets for a very short period, but these were replaced by Gladiators. Later as Gauntlets were replaced by Gladiators, Hurricanes and Spitfires, they were transferred to Auxiliary Air Force (AAF) squadrons including 504, 601, 602, 615 and 616. The latter squadron, with the Earl of Lincoln as CO, was the only home-based Gauntlet unit to operate after the outbreak of the Second World War, albeit for only a month – they were replaced by Spitfires in October 1939.

Until the 1938 Munich crisis, Gauntlets flew in overall silver finish bedecked with the ostentatious squadron markings of between-the-wars RAF fighter units. However, with the rapidly worsening political situation in Europe during 1938, Gauntlets were painted in standard RAF camouflage, although the Gauntlets of 79 Squadron were an exception as their role of a night fighter unit warranted them an overall matt black finish.

Overseas RAF Gauntlets served with 6, 33, 112 Squadrons and 1414 Met Flight (weather reconnaissance) at Eastleigh, Nairobi, which took four Gauntlets on charge as late as 1943. Apart from co-operating with ground forces in Palestine against tribes of dissident Arabs, Gauntlets saw action in the Middle East during the first fifteen months of the Second World War. In August and September 1940, when operated by 430 Flight, they joined with Vickers Vincent biplanes in attacking the Italians at Fort Galabat and Metemma airfield. On 7 September an Italian Caproni Ca 133 three-engined transport plane was shot down by Flt Lt A.B. Mitchell piloting Gauntlet K5355. The last recorded action by RAF Gauntlets was on 5 October 1940, when two (K5355 and K4295) flew escort to Vickers Wellesley bombers on a raid against Calabat, where the Gauntlets also dropped 20 lb Cooper bombs. Six Middle East

This immaculate Gloster Gauntlet II has been beautifully restored and exhibited by the Finnish Air Force Museum. It was one of twenty-five ex-RAF Gauntlets supplied to the Finnish Air Force early in 1940. (Courtesy Roger P. Wasley)

Gauntlets were transferred to the RAAF during September and October 1940. They flew with 3 Squadron on training before taking action in support of General Wavell's first Libyan campaign on 9 December 1940. Dive-bombing missions were flown in the opening phase of the attack, five raids being made on Italian transport at Sofafi on the first day. Air cover was provided for the Western Desert Force, and on 11 December 3 Squadron's Gauntlets dive-bombed a retreating enemy force which had been cut off at Sofafi by the 7th Armoured Division. The next day similar attacks were made by the Australian Gauntlets. Nevertheless all Gauntlets were suddenly withdrawn from the battle and replaced by Gloster Gladiators. The reason given for this hasty dismissal of the Gauntlets was lack of spares and high costs!

Meanwhile the Danish Government had signed to licence-build seventeen Gauntlets in Denmark after a pattern aircraft had been purchased. Danish Gauntlets flew with No. 1 Eskadrille at Vaerlose, near Copenhagen, and were still in service when the Germans attacked Denmark on 9 April 1940.

Early in 1940 twenty-five Gauntlets transferred from the RAF to the Finnish Air Force after the Soviet assault on Finland, which had opened in November 1939. One of these Gauntlets has been restored and is an exhibit at the Finnish Air Museum. Other Gauntlets operated with the South African Air Force, and a small number were in Southern Rhodesia with the Air Section of the Permanent Staff at Salisbury.

GLOSTER FS/36 (TSR38)

During November 1930 design started at GAC on the Gloster FS/36 in response to Air Ministry Specification S9/30 calling for a three-seat, torpedo-spotter reconnaissance aircraft to serve with the Fleet Air Arm. Another contender was Fairey Aviation, its design later evolving into the famous Swordfish under Specification 15/33, which also produced Blackburn Aircraft's well-known Shark.

Fairey's original S9/30 (S1706) had a Rolls-Royce Kestrel in-line engine, but its immediate

With a Rolls-Royce Goshawk steam-cooled engine, this was the one-off Gloster TSR38 torpedo-spotter-reconnaissance-type for naval use. Note the underwing bomb racks and spatted tailwheel of S1705. (GAC via Hawker Siddeley Aviation)

successors, the TSR1 and TSRII (Swordfish) had a Bristol Pegasus radial. Blackburn fitted an Armstrong Siddeley Tiger radial to its Shark, but at GAC Folland used a 600 hp Rolls-Royce Kestrel IIMS liquid-cooled 'V' for the company's FS/36 design. Constructed mainly from non-corrodible metal, with fabric covering except for the cowling, Gloster's FS/36 was a single-bay biplane carrying a three-man crew in tandem open cockpits. In order to ensure conveyance of an 18 in aerial torpedo beneath the fuselage, Folland designed a sturdy divided landing gear, and main wheels with heavy duty tyres. The tailwheel was partly enclosed by a streamline fairing.

Numbered S1705, Gloster's FS/36 made its initial flight from Brockworth in April 1932, trials continuing until the aircraft was sent to Martlesham for official type testing. Criticism was levelled at manoeuvrability and poor control response below 90 mph, but a top speed of 157 mph was recorded, and at speeds between 90 and 115 mph a marked improvement in handling characteristics was noticed. Favourable comments were made with regard to the pilot's excellent field of all-round view.

After returning to Hucclecote S1705 was modified in response to Specification 15/33 requiring a general purpose torpedo-spotter reconnaissance type. In its updated form S1705 emerged at Hucclecote in the summer of 1933 as the TSR38. Armament comprised two .303 in machine-guns, one firing forward and one in the rear cockpit, and an alternative to the 18 in Mk VIII torpedo was one 500 lb and two 250 lb, or six 250 lb bombs carried beneath the fuselage and wings. Folland later exchanged the Kestrel engine for a supercharged and steam-cooled 690 hp Rolls-Royce Goshawk VIII. After despatch to Martlesham for official tests, S1705 later carried out sea trials aboard the aircraft-carrier HMS *Courageous*.

A return to the A&AEE for further trials resulted in a top speed of 152 mph and a climb to 10,000 ft in under ten minutes. Despite this performance Gloster's TSR38 was not accepted by the Air Ministry, and one wonders if a radial engine, and not a steam-cooled Goshawk, had been installed would Gloster's torpedo bomber have had a more promising future?

Gloster F5/34 Unnamed Fighter

The F5/34 was Gloster's first monoplane fighter, responding to an Air Ministry specification which required a single-seat interceptor armed with eight .303 in machine-guns and equipped with retractable landing gear, enclosed cockpit, pilot's oxygen supply, a top speed of around 275 mph at 15,000 ft and a service ceiling of 33,000 ft.

Folland began work on the F5/34 (his last design for GAC) early in 1935. The result was a clean, low-wing monoplane of all-metal stressed-skin construction, fabric-covered movable surfaces, hydraulically operated split flaps, and a Dowty landing gear which retracted backwards and upwards into the wing. Part of each main wheel was exposed so that, in the event of a forced landing, they would bear the aircraft's weight! The tailwheel was designed to retract and respond similarly. Access to eight wing-mounted .303 in Browning machine-guns and rapid re-armament between missions was facilitated by hinged panels on the top wing surfaces. The 840 hp Bristol Mercury IX radial engine installed drove a three-blade de Havilland propeller.

The initial flight of the first F5/34 (K5604) was made from Brockworth in December 1936, with GAC chief test pilot P.E.G. Sayer at the controls. The new fighter's performance compared favourably with the Hawker Hurricane Mk I, but delays at Hucclecote, where Gladiator production was taking priority, meant the second F5/34 (K8089) did not fly until March 1938.

With a performance almost equalling the Hurricane, this was the Gloster F5/34 Unnamed Fighter armed with eight Browning machine-guns in the wings. (Rolls-Royce plc)

By then Hurricanes and Spitfires were in production for the RAF as Great Britain prepared to combat the ominous threat of Nazi Germany, and no production orders were issued for the Gloster F5/34 Unnamed Fighter. One feature which may have caused some doubt about its acceptance was its one-piece wing which, if damaged in combat, would put a complete aircraft out of action. In the case of bolted-on-type wings, as fitted to Hurricanes and Spitfires, if similarly damaged only a port or starboard wing replacement would be necessary.

GLADIATOR

As the last biplane fighter to enter first line service with the RAF, Gloster's Gladiator saw valiant service in the early years of the Second World War, and was to become a legend in its own right.

Developed by Henry Folland as a private venture from his Gauntlet, the prototype Gladiator (G-37), first known as the SS37, made its initial flight from Brockworth on 12 September 1934 piloted by P.E.G. Sayer. The machine conformed to a very demanding Specification F7/30 which had been issued by the Air Ministry in 1930, but GAC's pre-occupation with Gauntlet development had resulted in the company taking only a cursory interest in F7/30 requirements. However, when Folland and his team began a programme of updating for the Gauntlet in 1933, they decided that with airframe refinements, delivery later of a promised Bristol Mercury engine producing 800 hp, and an extra pair of machine-guns, their revised fighter could enter the F7/30 competition. Six other manufacturers had submitted designs, but none was very successful due mainly to the Air Ministry's choice of a steam-cooled Rolls-Royce Goshawk engine.

At GAC Folland retained some Gauntlet features, but fuselage longerons and stringers were modified, a large surface-type oil cooler fitted, and a faired headrest added for the pilot. The Gauntlet-style empennage and spatted tailwheel was retained. A 645 hp Bristol Mercury VIS radial engine was intended to provide the power, but problems with this engine resulted in a Mercury IV being installed within a Gauntlet-type cowling.

The SS37 possessed single-bay wings and simplified rigging wires in comparison to the two-bay Gauntlet, its new Dowty landing gear with cantilever legs adding to the drag-reducing properties. Indeed, the revised wing configuration and landing gear increased the new fighter's speed by some 15 mph with the Mercury IV engine, and when a Mercury VIS was installed the top speed was 242 mph at 11,500 ft with full four-gun armament.

After trials at Hucclecote and the A&AEE, G-37 (serial number K5200) passed to the Air Ministry on 3 April 1935 for its evaluation trials. It was back at Hucclecote on 2 July, and two days later appeared in the New Types Park at the annual Hendon show. On returning to GAC K5200 was updated to Gladiator Mk I standard, a proposal from Gloster which had been approved by the Air Ministry in June for an improved SS37. Modifications included an 830 hp Mercury IX engine, fully enclosed cockpit with sliding hood, improved wheel discs, unspatted tailwheel and alterations to the tail unit. A two-blade Watts wooden propeller was fitted, as there was no advantage in providing a three-blade type at that time. Top speed was now 253 mph at 14,000 ft and with trailing edge split flaps on all wings the landing speed was 48 mph. In this form K5200 served with the RAF until 12 November 1942, when it was scrapped.

The Air Ministry ordered an initial production batch of twenty-three Gladiator Is to Specification F14/35, the first aircraft (K6129) being delivered to the RAF on 16 February 1937. A further 180 machines were ordered with a proviso that they must all be on RAF charge by the end of 1937. Gladiators first entered service with 72 Squadron at Church Fenton; the five machines (K6130–K6134) were collected from Hucclecote by squadron pilots and flown direct to their new base. Gladiators continued equipping RAF fighter squadrons, including 3 (Kenley),

25 (Hawkinge), 33 (Egypt), 54 (Hornchurch), 56 (North Weald), 65 (Hornchurch), 80 (Egypt), 85 and 87 (Debden), 94 (Aden Protectorate), 112 (Egypt), 125 (Acklington and Leconfield), 127 (Syria), 247 (Roborough), 261 (Malta), 263 (Filton) and 521 'Met' Squadron (Bircham Newton). Gladiators also flew with 603, 605, 607 and 615 Squadrons, AuxAF, and as Sea Gladiators with the Fleet Air Arm (FAA), when they were fitted with deck arrester hook plus an inflatable dinghy contained in a fairing beneath the fuselage.

Meanwhile several overseas air arms had shown considerable interest in Gloster's new fighter during 1937, and subsequently Latvia purchased 26 Gladiators, Lithuania 14 (some flown with Russian markings in 1940), Norway 12, Sweden 55 (No. 8 Fighter Wing fought alongside Finland against Russia), Belgium 22 and China 36 (fought against Japanese during the 'China Incident' and provided air defence of Siuchow in 1938).

Although obsolete by the time they saw action during the early phases of the Second World War, Gladiators fought in France with the AASF when 607 and 615 Squadrons accounted for a number of German aircraft. One of the first German planes to raid Great Britain was a Heinkel He 111 bomber, which was shot down over the Firth of Forth by a Gladiator of 603 Squadron. Three Gladiators from 152 Squadron (Acklington) were detached to Sumburgh (Shetlands) for convoy and interception duties, and provided main air protection for the flying-boat base at Sullom Voe. When it was attacked by Heinkel He 111 bombers in January 1940 the Gladiators promptly engaged the enemy, shooting one down and damaging another.

In August 1940 the Sumburgh flight moved south as a nucleus for 247 Squadron, reforming at Roborough with Gladiators to provide protection for Plymouth and the Royal Naval dockyard. Round-the-clock patrols were mounted, night sorties being flown by a detachment at St Eval. On the night of 28 October 1940 a St Eval Gladiator patrol attacked and damaged a Heinkel He 111, another Heinkel receiving similar treatment on 7 November.

Meanwhile a further order for twenty-eight Gladiator Is had been received by GAC during 1937 (L8005–L8032), all of which went to 27 MU from whence they were sent mostly to the Middle East (ME) – an area in which Gladiators played an important part during the early war years. The first ME Gladiators arrived at Aboukir in 1938 and went to 33 Squadron, which used its machines on policing duties over Cairo, Jerusalem and Egypt's western borders. After the Second World War erupted this squadron moved to Mersa Matruh in the Western Desert and, once Mussolini had joined Hitler on 10 June 1940, began taking on the Italians. During a strafing attack against Sidi Aziz they met a Caproni Ca 310 bomber escorted by three Fiat CR32 fighters; in the ensuing fight the bomber and one Fiat were shot down without loss to the Gladiators. A few days later there was a dog-fight between 33 Squadron, joined by one Hurricane (the only one in the ME at that time) and a dozen Fiats. Two Italian machines were shot down for the loss of one Gladiator. By the time 33 Squadron moved to Helwan, Egypt, in July 1940 they had shot down thirty-eight enemy aircraft and destroyed twenty others on the ground. The cost was four pilots and eight Gladiators.

When Mussolini attacked Greece the RAF sent five units to assist with defence of the Balkans; two Gladiator and three Bristol Blenheim squadrons. Gladiators of 80 Squadron gave a good account of themselves over Greece, and by the end of 1940 had destroyed forty enemy aircraft for the loss of one Gladiator. On 9 February 1941 the squadron met a large formation of Fiat CR42 biplane fighters and, in the inevitable dog-fight, four Fiats were shot down and three claimed as probables. Then on 28 February the Italians suffered a humiliating defeat when twelve Gladiators and sixteen Hurricanes from 33, 80 and 112 Squadrons pounced on a swarm of their aircraft. Out of twenty-seven enemy machines shot down, plus eleven probables, Gladiators accounted for eleven of those destroyed and six of the probables.

From 26 March 1939 the Gladiators of 94 Squadron were based in Aden, and on 23 June 1940

Gladiator Is for the Irish Air Corps, waiting at Hucclecote for delivery to Baldonnel in 1938. The white building in the background is the Victoria public house, located across the road (now the A417 trunk) opposite Gloster's factory. (Rolls-Royce plc)

an Italian SM81 bomber was shot down at Ra Imram. Another Gladiator assisted in the capture of the Italian submarine *Galilei Galileo* when she surfaced off Aden, and on 6 July six Gladiators from 94 Squadron attacked Massala aerodrome and left seven Fiat fighters burning on the ground. This squadron supported the British advance through Somaliland, which was to end in the downfall of Italy's rule in East Africa. During the Iraq crisis (an attempted pro-Axis coup in April 1940), Gladiators shot down one Heinkel He 111 bomber and two twin-engined Messerschmitt Bf 110 fighters for the loss of one Gladiator. The last ME Gladiator unit was 6 Squadron, which flew Mk II machines from Wadi Halfa, and had a detachment at Kufra for duties over south-east Libya.

No account of the Gladiator can omit reference to 'Faith', 'Hope' and 'Charity', three Sea Gladiators which helped defend Malta in June 1940, and which gave rise to the famous legend. This has been attributed to the contemporary Maltese press, which depicted three Gladiators fighting alone in a sky black with Fascist bombers. Factual accounts toned down the story; the sky was not black with enemy bombers and there were some Hawker Hurricane fighters to lend a hand. Nevertheless, three Gladiators were very much involved in the aerial conflict over Malta. They were part of a consignment of Sea Gladiators for Royal Navy use, six having been left in crates at Kalafrana seaplane base. After Italy entered the war, RAF personnel had Admiralty permission to assemble the Gladiators for use as Malta's local air defence. At the same time Hurricanes and Blenheims were passing through the island en route for the ME, although none was actually retained for Malta's defence. On 28 June, however, four Hurricanes did arrive from Egypt to complement the Gladiators.

On approaching Malta Italian bombers proved too fast for the Gladiators, so the biplanes charged enemy formations spoiling their attempts at accurate bombing. On 22 June Sqn Ldr Burges shot down an SM79 bomber with his Gladiator, but strain on the Gladiators' engines was by then intense. Two machines (N5519 and N5520) had improvised Mercury VIII engines installed which were intended for a Blenheim bomber! So Malta had its legend. In an island devoted to St Paul, it came from Corinthians – 'and now abideth Faith, Hope, Charity, these three'. As they watched those three Gladiators in the war-torn skies above, the people of Malta knew their faith assured them of a safe future.

In very different temperatures the gallant pilots of 263 Squadron struggled to keep their

High over its native Gloucestershire, hybrid and refurbished Gladiator, G-AMRK, later to become L8032 owned and flown by the Shuttleworth Trust at Old Warden. (Courtesy Russell Adams FRPS)

Gladiators flying in appalling sub-zero conditions during the 1940 Norwegian Campaign. From their base, the frozen Lake Lesjaskog, those on the first campaign in April fought against the Luftwaffe; the lake was eventually abandoned, the serviceable aircraft being destroyed before the unit left for home in a cargo ship. In May 263 Squadron returned to Norway aboard the carrier HMS *Glorious*. With Gladiator IIs they were based at Bardufoss, and were credited with thirty enemy aircraft destroyed (a total of fifty for both campaigns). Sadly the *Glorious* was sunk en route for home by the German battleships *Scharnhorst* and *Gneisenau*. In 1940 FAA carrier-borne Sea Gladiators also operated over the North Sea, in Norway (804 Squadron) and the Mediterranean, 813 Squadron's machines participating in the naval bombardment of Bardia in August.

There are a few extant Gladiators around: K8042 in the RAF Museum collection; N5641 (263 Squadron colours) in a Norwegian museum; Swedish J8A Gladiator (278) at the Swedish Air Force Museum; N5519 fuselage and propeller at the War Museum, Fort St Elmo, Malta; L8032 (the last flying Gladiator) owned and flown by the Shuttleworth Trust at Old Warden aerodrome, Bedfordshire, and N2276 (N5903) at the FAA Museum, Yeovilton.

GLOSTER'S HAWKER LEGACY

Once Hawker Aircraft had taken over Gloster's an expansion plan was put into effect, and a new works occupying a further 24 acres (No. 2 Shadow Factory) was erected at Brockworth. Thus by 1939 and the onset of the Second World War, one million square feet of production space was available for aircraft production on the Hucclecote and Brockworth sites. This all provided an enormous boost to employment in surrounding districts, with a number of new houses erected for GAC workers, plus a small estate of prefabricated concrete houses complete with cinema erected at Timberlands, Brockworth. These were demolished after the war and later replaced by the large Abbotswood estate.

As production increased more employees joined those already travelling to Brockworth from Gloucester, Cheltenham, Stroud, Forest of Dean and other Gloucestershire towns and villages.

Kearseys Coaches of Cheltenham and other bus companies ran fleets of single and double-decker buses to and from the works round the clock. Satellite works were established in the county to avoid a concentration of work being subjected to German air attacks. Among these were Bentham (experimental), Standard Match Co. (Hempsted), Bristol Omnibus bus depot (Gloucester), Moreton Valence (works and aerodrome), and smaller units at Regent Street and Winchcombe Street, Cheltenham. Counting all sites GAC was reckoned at its peak to be employing some 12,500 people, which can't have been bad for the local economy!

Inevitably GAC's main factory became a German target and in October 1941 a lone Junkers Ju 88 dropped several oil bombs, one of which hit the tool room. Five people were killed or died of their injuries later and fifty injured. On 4 April 1942 again a single Luftwaffe bomber hit the factory, this time at shift change-over time; one bomb dropped into the crowded car park killing thirteen and injuring many more. In addition some 200 private homes in the vicinity were damaged, but production of Hawker Hurricanes for the RAF was not disrupted.

Meanwhile Hawker Aircraft, having introduced their own construction techniques into the Gauntlet and later Gladiator, subcontracted a number of its own Kingston designs to be built in quantity at Hucclecote. The first order was for twenty-one Hawker Hardy general purpose biplanes (K4050–K4070) all of which went to 30 Squadron at Mosul, Iraq, in 1935. A further batch of sixteen Hardys (K4306–K4321), all Gloster-built, went mostly to 6 Squadron, Palestine, and a final batch of ten machines (K5914–K5923) followed in 1936.

Hardys were fitted with a tropical radiator, message pick-up hook, water containers, tropical survival kit and underwing attachments to carry 112 lb supply containers or bombs. It also featured balloon low pressure tyres and very long exhaust pipes. They flew policing duties in the Middle East and over the North West Frontier from 1935, and when the Second World War erupted the Hardys went to 237 (Rhodesia) Squadron in Kenya, from where they operated against Italian forces until 1941.

The Hawker Hart, of which the Hardy was a derivative, was a tandem two-seat day bomber biplane of fabric-covered all-metal construction. With a Rolls-Royce Kestrel engine it provided a basic airframe from which evolved several off-shoots. Arguably best of these was the Audax, an army co-operation machine of which seventy-two were initially to have been built by GAC, but this contract was altered for Gloster's to produce thirty-two Hart Trainers and forty Hart Specials. Hart Trainers differed in having their Scarff gun rings removed from the rear cockpit, dual controls fitted, a normal type windscreen in the second cockpit, braked landing gear with the wheels slightly more forward, lengthy exhaust pipes and a $2\frac{1}{2}°$ reduction in the upper wing sweepback. Those built at GAC were referred to as Hart Trainer Interim due to their conversion from the original Audax order. The residue of forty Hart Specials produced at Gloster's from this contract were basically Audaxes converted to Hart standard, but having desert equipment, tropical radiator, heavy duty tyres, braked landing gear and a de-rated 510 hp Kestrel X engine. All GAC's Hart Specials served with RAF ME, some in Egypt and the rest in Palestine with 6 Squadron.

Gloster's built twenty-five Audax Indias in response to Air Ministry requirements for an aircraft specially suited to RAF service in India. The Audax was ideal for this purpose and an order placed with GAC starting with K4838, this machine remaining in the UK. The remaining twenty-four were delivered mostly to 20 Squadron, with 60 Squadron taking three temporarily on charge and K4853 going to the RAF Delhi Communications Flight.

The aforementioned expansion programme was by now under way, and work on the new shadow factory at Brockworth commenced in August 1938. It was completed in November 1940, the new site covering a total of 43 acres, while a lengthy concrete runway was laid down to cope with the increasing quantity of aircraft being produced in the two factories. Indeed, between October 1938

This 1942 aerial view shows No. 2 Shadow Factory, Brockworth, the airfield, with its relatively new concrete runway, and beyond that Gloster Aircraft's Hucclecote works and offices. (Author's collection)

and September 1940, as construction of Hurricane fighters increased on site, Gloster's produced the complete production batch of 200 Hawker Henley TTIII target-tugs (L3243–L3442).

Originally intended as a light bomber, the Henley monoplane was stressed for dive-bombing, the prototype (K5115) making its first flight on 10 March 1937. After successful trials a career as a RAF close-support aircraft seemed all set for the Henley, but Air Ministry policy on dive-bombing and ground attack aircraft changed. There was also an acute shortage of constant-speed propellers through demand for Hurricanes and Spitfires. This finished the Henley's dive-bombing/close-support concept; instead it was considered an ideal target-tug for the faster fighters which were then entering service. Hawker's promptly flew the second prototype Henley (7554) to Brockworth, where it was subcontracted to GAC for conversion as the TTIII target-tug.

First production Hawker Hardy (K4050) on Brockworth airfield in 1934. It later served in Iraq with 30 Squadron, RAF. (H.G. Hawker Engineering Co. Ltd)

Gloster-built Hawker Henley TTIII, L3243, shows off its clean lines above the clouds in 1939. (GAC)

An Air Ministry order followed for 200 production machines, also subcontracted to GAC.

Gloster's heavy involvement with quantity production of the Hawker Hurricane saw the first GAC-built machine (P2535) being rolled out on 27 October 1939. Peak production at GAC reached 130 a month, and by the end of 1940 as many as 1,000 Mk I Hurricanes had been built for the RAF. On early machines fabric covering was used over the metal wing structure, but subsequent models had a stressed-skin metal covering. Power was provided by a 1,030 hp Rolls-Royce Merlin II engine, although Gloster's first 1,819 Hurricanes had a Merlin III with a standard shaft capable of taking either a de Havilland or Rotol constant-speed propeller.

Altogether GAC produced 2,750 Hurricanes (Mk I, IA, IIA and IIB), the later machines

HM King George VI and Queen Elizabeth (later Queen Mother) on a visit to Gloster's Hurricane production line at Hucclecote in 1940. Forefront (next to the bench) is GAC general manager, Frank McKenna. Third row back, centre, can be discerned Lord Portal, then Chief of the Air Staff. (Author's collection)

Gloster-built Hurricane IIA (V...8/XR-J) early in 1941, when with 71 'Eagle' Squadron, RAF. (BAe plc)

A Hawker Typhoon IB in the capable hands of GAC production staff at Brockworth, c. 1943. (Author's collection)

Gloster-built Typhoon IB, with early style cockpit canopy, radio mast and three-blade propeller. The later version had a 'bubble' hood, whip aerial and four-blade propeller. (Hawker Siddeley Aviation)

having uprated Merlin XX engines. Some GAC-built Hurricanes flew in the Battle of Britain in 1940, others with RAF units in the ME (fitted with Vokes air filter), and a number went to the Soviet Red Air Force. Quite a few were converted to Sea Hurricanes, including several 'Hurricats' for catapulting off merchant ships to chase off Focke-Wulf Condors menacing Allied convoys, but the majority were FAA carrier-borne fighters fitted with a deck arrester hook.

The final Hawker-type built at Gloster's was the Typhoon which, despite early problems with its Napier Sabre engine and failure in the interceptor role, became one of the Second World War's most aggressive close-support aircraft. Of all-metal construction Typhoons had progressively more powerful Napier Sabre engines fitted, from 2,000 to 2,260 hp, improved cockpit hood, four-blade propeller, whip aerial and other refinements. Armed with four 20 mm cannon, bombs or underwing rockets, Typhoons earned a fearsome reputation over occupied Europe. They were known as the 'Scourge of the Panzers', sweeping in from the sea at zero feet, flattening everything in their path and leaving a trail of carnage. After destroying German radar installations in June 1944 prior to D-Day, RAF (2nd Tactical Air Force) Typhoons decimated concentrations of German armour ahead of Avranches and in the Falais Gap during the Battle of the Bulge.

There were 3,317 Hawker Typhoons built and, apart from two prototypes and the first fifteen production aircraft, 3,300 of them rolled off Gloster's Brockworth production lines. Yet with the cessation of hostilities, like the violent storm they were named after, Typhoons vanished quickly from the scene.

GLOSTER F9/37

British twin-engined fighters of the Second World War included the Bristol Blenheim IF and Beaufighter, de Havilland's Mosquito fighter variants, and Westland's Whirlwind and Welkin. Gloster's produced the lesser known F9/37 which, but for an unfortunate accident, could well have turned the tide in Great Britain's favour early on in the war.

The design was based on an earlier GAC project responding to Specification F5/33 which had called for a two-seat turret fighter. Together with Armstrong Whitworth and Bristol entries this was rejected, Gloster having envisaged a twin-engine type with four-gun dorsal turret and 625 hp Bristol Aquila radials. A later Specification (F34/35) for a revised two-seat turret fighter led to a second Gloster design with four-gun turret and fixed armament in the nose. Serial K8625 was allotted, but acceptance of Boulton & Paul's Defiant for RAF service led to cancellation of Gloster's project. When Specification F9/37 was issued eighteen months later, calling for a single-seat twin-engine fighter with fixed forward-firing armament, GAC chief designer W.G. Carter submitted a layout based on Gloster's earlier F34/35 design.

The dorsal turret was deleted, cockpit slightly elevated, nose section altered and amendments carried out to the twin fins and rudders profile. The cockpit canopy was refashioned and further glazing added just aft of the main canopy. Engines were 1,050 hp Bristol Taurus TE/1, 14-cylinder two-row radials driving three-blade Rotol variable-pitch propellers. The all-metal airframe had stressed-skin metal covering with fabric-covered movable controls. A detachable nose cone provided access to the heater, instrument panel, rudder pedals and camera-gun, while the extreme rear fuselage, also detachable, contained a navigation light and gave access to the tailwheel.

Intended for assembly by subcontractors to GAC, the F9/37 fuselage consisted of a front portion containing the cockpit and two 20 mm nose cannon, and a rear section comprising two sub-assemblies, one forward with four .303 in Browning guns or three 20 mm Hispano cannon (plus wing attachment points), and a rearmost section with tailplane attachment fittings and the retractable tailwheel mounting. Wing construction was mainly Duralumin, with a centre-section

Second prototype Gloster F9/37 (L8002), with Rolls-Royce Peregrine engines. Note the nose-mounted guns. (Author's collection)

containing fuel tanks and engine bearers, and outer panels. The landing gear main wheels retracted backwards into the engine nacelles.

GAC received a contract for two prototype F9/37s (L7999 and L8002), the first making its initial flight from Brockworth on 3 April 1939 piloted by P.E.G. Sayer. No gun armament was then fitted and blanking plates covered the nose position gun ports. On 23 May L7999 gave a spectacular performance at Northolt in front of a party of VIPs, its performance surprising some MPs. This aircraft was sent to the A&AEE for its official trials on 8 July, and the Martlesham pilots gave it excellent reports with regard to handling, comfort and good forward view, although port and starboard views were limited by the engine nacelles. Performance was good with a top speed of 360 mph at 15,000 ft and Gloster's F9/37 looked a winner. In July 1939, however, while still at Martlesham, L7999 was badly damaged in a landing accident and returned to Hucclecote for major repairs. By then GAC was heavily engaged in Hurricane production and could not return L7999 to the A&AEE until April 1940. The engines now fitted were 900 hp Taurus T-S (a) III in which the superchargers had been derated and the top speed reduced to 323 mph at 15,000 ft.

The second prototype (L8002), powered by two 885 hp Rolls-Royce Peregrine engines, was armed with two 20 mm Hispano cannon in the nose and an unorthodox location of either four .303 in Browning guns, or three 20 mm cannon, in the top of the fuselage aft of the cockpit. These guns fired above the pilot's head, and to allow for this the fuselage was slightly tapered aft of the cockpit. In 1940 this formidable armament made the F9/37 the world's most heavily armed single-seat fighter, and had it not been for L7999's unfortunate crash in July 1939, which retarded the type's progress by twelve months, Gloster's twin-engined fighter might well have changed the course of history at a time when Britain so desperately needed a heavily armed long-range fighter.

E28/39 AND E1/44 – GLOSTER'S SINGLE-ENGINED JETS

In March 1936 a new company named Power Jets was formed by a group of businessmen who were backing the innovative ideas of a young RAF officer named Frank Whittle (later Sir Frank), who had designed an aero-engine in which gas turbines and jet propulsion were combined in one unit – the turbojet.

Despite official scepticism of his plans, Whittle's first jet engine, the 'U', made its initial test-bench run in April 1937. Initially it overran out of control, but within twelve months modifications showed positive results. The Air Ministry signed a contract for jet-engine development with Power Jets on 18 March 1938, and work began on a new engine, the W1X built for Power Jets by British Thomson Houston. On 7 July 1939 the Air Ministry ordered a second engine for flight testing to be known as the W1. A suitable airframe was now required in which this new type of power unit could be installed, and Whittle visited Hucclecote to discuss proposals with GAC chief designer George Carter. In preference to adapting an original Gloster piston-engined fighter design for jet propulsion (F18/37), Carter designed a completely new layout for the revolutionary engine. By 3 February 1940 the basic design of the new aircraft was complete and the Air Ministry ordered two prototypes from Gloster's to Specification E28/39.

Although intended as a flying test-bed for Whittle's new jet engine, the design had to fulfil single-seat fighter requirements, and the speed required was about 380 mph given an engine thrust of 1,200 lb. The E28/39 was a small low-wing monoplane with a tricycle landing gear. The initial work on it took place at Hucclecote, but with Hurricane production making GAC a prime Luftwaffe target, the new jet was moved to Regent Motors in Cheltenham, where it was completed.

On 7 April 1941 the first machine (W4041) was transported back to Brockworth for taxying trials, with test pilot Gerry Sayer at the controls. Three 'hops' were made on the third day, but after returning to Cheltenham the new jet was moved to RAF Cranwell for official trials. With the W1 engine fitted its first flight was on 15 May, a day of great triumph for Whittle, Carter and the GAC team. Seven months later W4041 had a W1A turbojet fitted producing 860 lb of thrust, and tests resumed at Edgehill, Warwickshire. Problems arose with both turbine blades and oil viscosity at high altitude, and on 27 September 1942 the jet had to make a forced landing. Sadly Gerry Sayer was killed the following month testing a Typhoon with a new gunsight, and when E28/39 testing resumed in November, test pilot Michael Daunt took over the programme. After being sent to the RAE, Farnborough, for tests, W4041 returned to Bentham, GAC's design and experimental unit near Brockworth, and a new Power Jets W2/500 engine of 1,700 lb thrust was installed.

Meanwhile the second E28/39 (W4046) was fitted with a Rover W2B turbojet (Rover Cars took over production of Power Jets) and sent to Edgehill in February 1943 to be flown by test pilot John Grierson. On 30 July RAE pilot Sqn Ldr Davie had climbed W4046 to 37,000 ft,

Britain's first jet-propelled aircraft, the Gloster E28/39 powered by a Whittle W1 turbojet, c. 1941. (GAC)

Gloster E1/44 jet fighter (TX145) receiving attention at the A&AEE, Boscombe Down in 1948. (Courtesy Roger P. Wasley)

when the ailerons jammed causing the jet to crash. Fortunately Sqn Ldr Davie escaped by parachute, but in his descent from 33,000 ft he suffered from frostbite.

In the meantime W4041 had been fitted with a W2/500 power unit and was tested at Brockworth and Barford St John. On one flight John Grierson, wearing a pressure suit and inhaling pure oxygen for thirty minutes before take-off, reached an altitude of 42,170 ft. Afterwards it was decided the E28/39 should be preserved, and on 29 April 1946 W4041 was placed on permanent display in the Science Museum, South Kensington, where she stands as a symbol of all that was great in British aviation.

Lesser known perhaps was the E1/44, another single-seat, single-engined jet fighter from Gloster's which evolved from an original design of 1942. This was when there was doubt as to the availability of Rover and Halford turbojets for the new Gloster F9/40 twin-engined fighter. A single-engine design (E5/42) was temporarily favoured by the Ministry of Aircraft Production, but successful flights of the F9/40 caused the single-jet idea to be cancelled. Gloster's, however, continued experimenting with the E5/42 design, and by November 1943 work on two updated machines each to be powered by a Rolls-Royce Nene turbojet had started at Bentham. An Air Ministry revival of Specification E5/42 as E1/44 caused many alterations and modifications to be carried out to the design, and it was not until late 1944 that a third airframe (SM809) was started. By 1947 this was ready for flight testing, but en route to the A&AEE, Boscombe Down, the aircraft's road transporter crashed on a steep hill, wrecking SM809 beyond repair.

The second E1/44 (TX145) was speeded up and made its first flight in the hands of GAC test pilot Bill Waterton on 9 March 1948. It achieved 620 mph at sea level, but poor handling qualities necessitated modifications to the tailplane. This warranted a third prototype (TX148), which first flew at Boscombe Down in 1949. Despite satisfactory performance figures the type did not reach production status, and work on a fourth prototype (TX150) was not completed. The other two prototypes were used at Farnborough on various brake-parachute and flight control system trials until withdrawal in the early 1950s.

ALBEMARLE

As Gloster's continued with wartime jet designs and development, an entire production batch of 600 Armstrong Whitworth Albemarle twin-engined transport/glider-tug aircraft was assembled by A.W. Hawkesley Ltd (another part of the Hawker Siddeley Group) in No. 2 Shadow Factory at Brockworth. The Albemarle, originally the Bristol Type 155 bomber to Specification P9/38, was passed to Armstrong Whitworth at Coventry as the AW41 Albemarle.

The first few were built as bombers, but an Air Ministry order required the type be produced as a glider-tug and special transport for the airborne forces. Albemarle prototype (P1360) made its initial flight on 20 March 1940, a main feature being its composite metal and wood construction, which not only saved the use of strategic materials, but meant a wide distribution of subcontracting could be undertaken by small non-aviation companies. These sub-assemblies were transported to Brockworth where Albemarles were completed, test-flown and delivered, powered by two 1,590 hp Bristol Hercules XI engines.

Of the 600 Albemarles built, the RAF received 99 Special Transports Mk I, 99 Mk II, 49 Mk V and 133 Mk VI respectively, plus 69 Glider-tugs Mk I, 117 Mk VI and some of the original bombers later converted accordingly. Ten Albemarles went to the Soviet Air Force, an RAF training unit being specially set up at Errol, Scotland, to familiarize Russian aircrew who were to ferry the Albemarles to Russia. Albemarle production ended at Brockworth in December 1944.

Albemarles saw their first action in July 1943 when, during the invasion of Sicily, they operated as glider-tugs with 296 and 297 Squadrons. On D-Day, 6 June 1944, six Albemarles flew as pathfinders for the 6th Airborne Division and dropped men of the 22nd Independent Parachute Company. Four Albemarle units later towed Horsa heavy transport gliders to France carrying more British paratroopers, and that September Albemarles from 296 and 297 Squadrons towed 1st Airborne Division gliders to Arnhem. The last Albemarles to retire were replaced by Halifaxes in February 1946.

A line-up of factory-fresh Armstrong Whitworth Albemarles at Brockworth in 1943. All 600 production machines were built by A.W. Hawkesley in No. 2 Shadow Factory, and flew as troop transports and glider tugs. (Courtesy Mike Charles)

Meteor

Seventeen months before Gloster's E28/39 jet aircraft made its first flight, chief designer George Carter had started work on a twin-jet fighter to Air Ministry Specification F9/40. Eight prototypes emerged from GAC, powered in turn by a variety of turbojet engines: the Rover/Power Jets W2B/23, Power Jets W2/500, Metropolitan Vickers MFV2, Rolls-Royce W2B/37 and de Havilland Halford H1s. The first flight of an F9/40 prototype took place on 5 March 1943, fifth machine (DG206) doing the honours piloted by GAC test pilot Michael Daunt, reaching a maximum speed of 430 mph. It was powered by two 2,300 lb thrust de Havilland H1s.

Structurally the F9/40, like succeeding Meteors, featured an all-metal airframe with stressed-skin covering. The fuselage was built in three portions, the front containing nosewheel, cockpit and armament, while engines, landing gear, main fuel tank (with provision for ventral auxiliary tank), flaps and air brakes were incorporated in the centre-section. The rear portion included the lower part of the fin built integrally. Wings were formed with alloy and steel spars, alloy ribs and push-rod-operated ailerons. Empennage featured a one-piece tailplane, balanced elevators, top fin and upper and lower rudder.

An initial batch of twenty F9/40s (officially named Meteor) was produced, the first (EE210) being flown in January 1944. Deliveries to the RAF began in July that year, with 616 Squadron receiving Meteor Mk Is powered by two Rolls-Royce Welland turbojets. Armament comprised four 20 mm Hispano cannon in the nose. Seven machines went to Manston, Kent, operating against German V1 flying bombs; on 4 August Flg Off Dean destroyed a V1 by turning the missile over with his Meteor's wing-tip after the guns jammed.

A proposed Mk 2 Meteor with Halford engines came to nothing, but the Mk 3 was a successful production version of which 210 were built, initially with Welland engines, replaced in later batches by Rolls-Royce Derwents. Meteor 3s arrived with 616 Squadron in December 1944, one flight joining the 2nd Tactical Air Force in January 1945. That April, 504 Squadron received Meteor 3s and this mark went on to serve with another thirteen RAF squadrons during 1945 and 1946.

All F9/40 and Meteor test flying by GAC was carried out at Moreton Valence, originally Haresfield aerodrome which opened in November 1939 as a training landing ground, and initially used by Staverton's 6 AONS Ansons. Rebuilt as Moreton Valence in 1941, with three runways and two hangars, it continued as a training establishment. However, in October 1943 GAC moved in with its FTD, by which time extra aircraft assembly sheds and a runway extension had been added, and thereafter extensive test flying of Meteors was concentrated at Moreton Valence. Some RAF Anson and Oxford trainers still used the airfield, but after VE-Day 83 Gliding School soon provided the only RAF presence and it too left in October 1946, leaving Gloster's the sole occupants. As the heavy programme of Meteor testing continued, a further extension to the runway was added in readiness for the arrival of the Gloster Javelin later.

Meanwhile several early Meteors had been used in various research programmes. The first Mk I (EE210) was sent to the USA in exchange for a Bell XP-59 Airacomet, an American jet which flew from Moreton Valence in September 1943. Sixth Meteor I (EE215) became the world's first jet aircraft to undergo reheat (afterburning) trials, while another machine (EE227) was fitted with Rolls-Royce Trent turbo-prop engines to become the world's first turbo-prop aircraft.

On 17 July 1945 the first Meteor Mk 4 flew with uprated Derwent engines in long nacelles and a strengthened airframe, and from the ninth Mk 4 built the wing-tips were clipped by 34 in which improved the roll rate, but increased landing and take-off speeds. Two Meteor 3s (EE454 and EE455) were updated to Mk 4 standard minus armament, ventral tank, VHF mast and given a high-gloss finish, one being in an overall yellow scheme (the 'Yellow Peril') in preparation for an attempt on the World Air Speed Record. Gp Capt Wilson in EE454 *Britannia* set up a new record

This was the 'Yellow Peril' (EE455), a Gloster Meteor F3 updated to F4 standard, and flown at 603 mph on 7 November 1945 by GAC test pilot Eric Greenwood. (GAC)

of 606 mph on 7 November 1945, GAC test pilot Eric Greenwood achieving 603 mph in EE455. On 7 September 1946 Gp Capt E.M. Donaldson flew Meteor 4 (EE549) at 616 mph. As well as RAF service, Meteor 4s flew with the air forces of Argentina, Belgium, Denmark, Egypt, France and the Netherlands. They also served as test-beds for a number of Rolls-Royce and other turbojets.

One Meteor 5 (VT347) was built as a reconnaissance Mk 4 with fuselage and nose-mounted cameras, but this machine crashed on its first flight from Moreton Valence on 15 June 1949 killing test pilot Rodney Dryland. A projected Mk 6 with Derwent 7s and revised tail unit was not built, but a prototype Mk 7 (G-AKPK) and 640 production aircraft were built to Specification T1/47 as T7 Meteor Trainers. This variant had a 30 in extension to the nose, which allowed a second seat in tandem to be fitted beneath a side-hinging canopy. Armament was deleted, a full dual-control system installed, and a 180-gallon ventral fuel tank fitted as standard.

Practically every RAF fighter unit had a Meteor T7 attached to it and they served with the CFS, FTSs, AFSs, OCUs and two units of the FAA. This version was also exported to Belgium, Brazil, Denmark, Egypt, France, Israel, the Netherlands, Sweden and Syria. Towards the end of T7 production uprated Derwent 8s were installed plus a Mk 8-type tail unit.

By the end of 1947 Meteors were losing their lead to more modern fighter designs. Thus to obtain the highest performance possible from a Meteor airframe with minimum modifications, Gloster's installed a pair of 3,600 lb thrust Derwent 8s, fitted straight tapered vertical tail surfaces, widened engine nacelle intakes, lengthened the fuselage, and fitted a 'bubble'-type cockpit canopy. Also included was an additional 95-gallon fuel tank, a Martin Baker ejector seat and a gyro gunsight. Designated Meteor 8, this variant's top speed was 598 mph at 10,000 ft.

Over 1,800 Meteor Mk 8s were built, serving with more than thirty RAF squadrons. Indeed, they were the mainstay of RAF Fighter Command during the 1950s, and also operated in the Korean War with 77 Squadron of the RAAF, when they tangled with Russian-built MiG-15 swept-wing jets. Meteor Mk 8s were sold abroad to Brazil, Denmark, Egypt, Israel, the Netherlands and Syria, as well as being built under licence in Belgium by Avions Fairey and by Fokker in Holland. At home Meteor Mk 8s were used for various research purposes with rocket and turbojet engines, and one machine (WK935) was fitted with a long nose to accommodate a prone pilot and was first flown on 10 February 1954. Quite a number of Mk 8s finished their days as brightly coloured pilotless radio-controlled target drones (designated U16 and U21) at missile ranges. One private venture Meteor Mk 8 was built as a ground-attack variant and named Reaper. It was armed with either twenty-four 60 lb rockets or two 1,000 lb bombs.

Gloster Meteor T7s and F8s taking shape at GAC, Hucclecote, in the 1950s. (GAC)

This smart looking Meteor T7 (WA662) is seen operating with the RAE, Llanbedr, c. 1960s. (Courtesy Roger P. Wasley)

The Meteor Mk 9 (FR9) was a fighter reconnaissance version with gun armament and a nose-mounted camera, while the PR10 was a high-altitude photographic reconnaissance-type, with cameras mounted in the fuselage and nose. It featured the long-span wings of a Mk 3 with a Mk 4 tail unit, and the prototype first flew on 29 March 1950.

Meteors NF11, 12, 13 and 14 were night fighter versions built by Armstrong Whitworth at Coventry. With radar equipment housed in a lengthened nose, the original NF11 was an updated T7 in which the rear cockpit accommodated a navigator/radar operator and his apparatus. Guns were wing-mounted, and 335 production Meteor night fighters went to fourteen RAF units. The NF12 was fitted with American radar and powered by Derwent 9 engines; NF13 was a tropicalized NF11 variant; and the NF14 a more refined update with revised 'double bubble' canopy, longer nose and an automatic stabilizer for improved directional stability.

In this superb shot, Meteor F8, VZ460, one of the first production batch F8s, is in use as a trials machine for rocket projectiles and bomb-carrying underwing pylons. (Courtesy Russell Adams FRPS)

Javelin

On the afternoon of 26 November 1951 a strange shape appeared in Gloucestershire's skies as Gloster's new Javelin fighter made its initial flight. Piloted by GAC test pilot Sqn Ldr W.A. 'Bill' Waterton, this large machine created much local interest, not only because it was manufactured in Gloucestershire, but because of its extraordinary triangular (delta) configuration. True there were a few delta-wing-types flying like Avro's 707, Boulton & Paul's P111 and Fairey's FD1, but these were experimental and research aircraft (Avro's giant delta-wing Vulcan bomber did not fly until 30 August 1952), so Gloster's GA5 Javelin was unique in being the first twin-jet powered delta-wing fighter in the world.

Unorthodox in appearance, with its 52 ft triangular wing-span and 17 ft horizontal delta tailplane perched above a large slab-sided fin and rudder, the Javelin was manufactured from light alloy practically throughout with the exception of some steel edging. The fuselage was in four sections, the main forward and centre portions being joined during assembly to produce a very robust structure to which the main wing spar was fitted; this spar, leading-edge ribs and alloy skinning formed a torsion box which extended to the wing-tip.

The inner wing section incorporated flaps, air brakes, fuel tanks, main landing gear and the gun and ammunition bays with two 30 mm Aden cannon to each wing. The nosewheel of the Dowty tricycle landing gear retracted backwards beneath the forward fuselage. Located each side of the cockpit were air intakes and ducts for the two 7,000 lb thrust Armstrong Siddeley Sapphire turbojets. The cockpit canopy on the first four Javelin prototypes had a perspex front portion for the pilot, with the navigator enclosed in a metal section with two portholes. This proved an unsatisfactory design and the fifth prototype (WT836) appeared with a new fully glazed twin canopy.

Capable of radar-guided defence and interception in bad weather at night, the Javelin's trials and development flying took time and its toll of setbacks. The first came on 29 June 1952 when, during a high-speed run, the first prototype (WD804) suffered severe elevator flutter and both elevators broke away. Test pilot Bill Waterton, despite the danger and risk involved, managed to get the Javelin down, albeit at an unavoidably fast landing speed. The landing gear collapsed and WD804 was written off, but Sqn Ldr Waterton escaped from the wreck with the auto-observer intact, and for his bravery and exceptional devotion to duty was awarded a well-earned George Medal.

The second prototype Javelin (WD808) made its first flight from Moreton Valence on 20 August 1952, and the following month gave an impressive display of low flying at the

The first prototype Gloster Javelin (WD804) at busy Moreton Valence (note the Meteors in the background), from where its first flight was made on 26 November 1951. (Courtesy Russell Adams FRPS)

Sqn Ldr W.A. 'Bill' Waterton, GAC chief test pilot in the early 1950s, who was responsible for testing the prototype Javelin. (Courtesy Russell Adams FRPS)

Wg Cdr R.F. 'Dicky' Martin, chief test pilot at GAC on Javelins from March 1954 until 1960. (Courtesy Russell Adams FRPS)

Gloster Javelins being produced in the main erecting shop, Brockworth, 1950s. (Courtesy of the Gloucester Citizen)

SBAC Farnborough Air Show. Other prototypes followed: WT827 in March 1953, WT830 in January 1954 (the first Javelin to take off from Brockworth), and WT836 the following July, all participating in the Javelin development flying programme. Other test pilots soon involved were Geoff Worral, Brian Smith, Jan Zurakowski and ex-Blackburn pilot Peter Lawrence, who tragically lost his life in WD808 on 11 June 1953, when it crashed on stalling tests near Bristol.

In March 1954 Wg Cdr R.F. 'Dicky' Martin took over as GAC's chief test pilot, and on 22 July that year the first production Javelin (XA544) took off from Brockworth with 8,300 lb thrust Armstrong Siddeley Sapphire turbojets. It subsequently appeared in September at the SBAC Show, where it was joined by the third production machine (XA546). Unfortunately XA546 crashed in the Bristol Channel during October and, sadly, test pilot Flt Lt Ross from Farnborough perished with the aircraft.

Meanwhile a modified Javelin wing was developed in which an outer extension reduced thickness/chord ratio towards the wing-tip and gave a kinked look to the leading edge. This wing had been tested on the ill-fated WD808, re-introduced on the fourth prototype and standardized on all subsequent Javelins. This early series of Javelins experienced rudder buffeting, and this was counteracted by fitting what was termed a 'pen knib' fairing between the tail pipes.

Javelins first entered RAF service with 46 Squadron, Odiham, on 29 February 1956, when they superseded Meteor NF 12 and NF 14 night fighters. The squadron's Javelin 1s were later replaced by Mk 2s with American radar.

The Mk 3 Javelin (T3) was a two-seat training variant with a longer fuselage and raised rear cockpit canopy; Mk 4s introduced an all-flying tailplane in which the elevators acted as anti-balance tabs; Javelin 5s had increased fuel capacity in revised wing tanks and provision to carry four de Havilland Firestreak AAMs on underwing pylons; and the Mk 6 had American A122 radar installed within a modified nose radome.

The first much refined Mk 7 (XH704) flew initially on 9 November 1956, with two rows of vortex generators on the wings to reduce shock wave and airflow separation at high speed, deletion of the pen knib fairing in favour of a fuselage extension beyond the fin trailing edge, a downward angle to the jet pipe nozzles, and installation of 11,000 lb thrust Sapphire Sa 7 turbojets. The gross weight of Gloster's Javelin was now in excess of 40,000 lb, but with the introduction of the Mk 8 a number of Mk 7s went into storage at 5 MU, Kemble, pending updating.

Javelin 8 was the last actual production version and had Sapphire SA 7R engines of 12,300 lb thrust with reheat. It also had a Sperry autopilot, droop-wing leading edge, an extra row of vortex generators, American radar, and deletion of the two inboard Aden guns (Firestreak AAMs decided on as Javelin's main armament). In 1960 GAC began a major new modification programme in which Mk 7s were brought up to Mk 8 standard and designated Javelin Mk 9. Those known as the F(AW)9R had an in-flight refuelling probe fitted on the starboard side of the forward fuselage. The first F(AW)9 conversion flew on 6 May 1959, with first deliveries going to 25 Squadron RAF.

Javelins, like many Gloster-built aircraft before them, were Britain's defenders in the skies, serving from the late 1950s to mid-1960s. They operated with numerous RAF units, the last UK-

Early production Javelin F(AW)1 (XA552), with ventral fuel tanks, banking to port high over Gloucestershire. (Rolls-Royce plc)

A Javelin F(AW)8 getting away from a wet Brockworth runway in 1958. Note the extended jet pipes, kinked wing leading edge and vortex generators on upper wing surfaces. (GAC)

based Javelin unit being 23 Squadron at Leuchars, which gave up its Gloster machines for English Electric Lightnings in November 1964. Overseas the last Javelins in Germany were those of 11 Squadron, disbanded at Geilenkirchen in January 1966, but 29 Squadron flew Javelins in Zambia during the 1965 Rhodesian crisis, and later in Cyprus until summer 1967. From Singapore 60 and 64 Squadrons operated Javelins over Malaya as defence against the Indonesian Air Force. On 30 April 1968, 60 Squadron made a final fly-past over Singapore, ending not only the Javelin's career but over forty years of Gloster Aircraft history. A total of 435 Javelins was built (including some by Armstrong Whitworth), and at its Hucclecote and Brockworth factories GAC had produced nearly 6,000 aircraft of all types before the last Javelin flew out on 8 April 1960. The last Javelin to fly was XH897, a Mk 9 used for a number of years at the A&AEE, Boscombe Down. On 25 January 1975 it flew in final salute over Gloucester and Brockworth before flying into retirement at the Imperial War Museum, Duxford.

This almost ends Gloster Aircraft's saga, but it is worth noting that No. 2 Shadow Factory was used just after the war by A.W. Hawkesley to build prefabricated bungalows to help meet the acute post-war housing shortage, with some 18,000 being produced at Brockworth in four years. During the Korean War period this factory was converted to produce Armstrong Siddeley jet engines as Brockworth Engineering Co. was formed and recruited staff nationally. Bristol Siddeley Engines (later part of Rolls-Royce) had engine test-beds erected on the far west side of Brockworth airfield; these buildings remain in use today as private businesses.

Meanwhile this late influx of work and activity inevitably led to further housing development around Brockworth, and hopes still ran high at GAC with the thin-wing Javelin proposal. As the ill-fated F153D supersonic fighter project, this was abandoned after a 1957 Government Defence White Paper stopped any further F153D development plans. Thereafter a variety of non-aviation work, spread among several GAC associated companies, attempted survival, but to no avail. In April 1964 the GAC Hucclecote site was sold to Gloucester Trading Estates, and Brockworth was taken over by British Nylon Spinners (later ICI Fibres). After the last Javelins had left Moreton Valence that site too was sold and converted to a trading estate. Today Brockworth airfield itself is doomed to disappear beneath yet another of those inevitable business parks and housing developments which are swallowing up the English countryside.

CHAPTER 3
Bristol Aeroplane Company

During 1910 Sir George White, Bt., who having started as a solicitor's clerk rose to become chairman of the Bristol Tramway and Carriage Co., turned his attention to the new field of aviation. This was no overnight decision as Sir George had been carefully observing the steady growth of early aeronautics since 1904, and was convinced that aeroplanes held the key to the future of long-distance transport. Visualizing aviation as Bristol's growth industry, Sir George White informed the annual meeting of the Tramway Co., held on 16 February 1910, of his intention to open an aircraft factory. Three days later the British and Colonial Aeroplane Co. Ltd was registered with a capital of £25,000. Within three months an aircraft works had opened using two sheds of the Tramway terminus at Filton, a Gloucestershire village four miles from Bristol.

Plans to produce the French Zodiac aeroplane under licence fell through, and the next design, which was based on the French Farman brothers' biplane ideas, evolved as the Bristol Boxkite. This went into production, and in 1911 Sir George White justifiably claimed that the world's largest aircraft factory was located at Filton. Land to the north of the two sheds was acquired as an airfield, more buildings erected to cope with expansion and Filton House purchased for offices. After war was declared in August 1914, Filton works was greatly extended, and by the Armistice in November 1918 more than 3,000 employees were on the payroll and over 2,000

Sir George White, Bt, founder of the British & Colonial Aeroplane Co. Ltd (later Bristol Aeroplane Co.) in 1910. (Courtesy the present Sir George White)

aeroplanes had been produced for military use. Filton airfield was used by the RFC from December 1915 in the preparation of new squadrons destined for France and the Western Front. A number of large hangars were also built for the South West Aircraft Acceptance Park, which was established at Filton for processing aircraft completed by local firms.

On 31 December 1919 the British and Colonial Aeroplane Co. officially became The Bristol Aeroplane Co. Ltd for clarity (hereafter referred to as Bristol or Bristol's). By then the aftermath of war had seen a great reduction of work at Filton, but the airfield remained active, and on 15 May 1923 it became an RAF RFS under a contract signed with Bristol. Several types of Bristol aircraft were used for instruction until de Havilland Tiger Moths became standard trainers in 1933. In the meantime 501 (County of Gloucester) Squadron had formed at Filton on 14 June 1929 as a special reserve unit flying Avro 504Ns, and later DH9As.

During 1930 work began at Filton on alterations to existing buildings, with new accommodation being provided for RAF personnel on a site north of the airfield. In 1937 the RFS became 2 E&RFTS, while Bristol's factory stepped up production rapidly as the RAF adopted the Blenheim as its standard light bomber. To cope with the increasing number of machines rolling out, it was necessary to form a Filton Ferry Flight (later 2 Ferry Pilot's Pool). As for 501 Squadron it remained at Filton equipped with Hurricane fighters; the unit's first scramble in the Second World War was on 11 November, but no enemy aircraft was sighted. 501 moved to Tangmere when 263 Squadron replaced it with its Gloster Gladiators, but this unit left for Norway in April 1940, which left Filton with only balloon barrage defences.

On 25 September 1940 the Luftwaffe made a heavy raid on Bristol's aircraft and aero-engine works, dropping 100 tons of bombs on the area and killing or injuring 238 people in the factories and 107 of the local populace. Next day Hurricanes of 504 Squadron arrived at Filton and, from another incoming German raid intercepted by Fighter Command over Dorset and Somerset, they shot down three Luftwaffe Bf 110s that managed to reach Bristol. German night attacks during August caused further damage and casualties, but interruption of work was lessened after Bristol's introduced their own warning system.

Bristol's Filton works from the air in 1921. (BAe plc)

Bristol Bulldog fighters for the RAAF being produced at Filton in 1929. Note the Type 110A five-seat commercial biplane (first flight 25 October 1929) at the end of the row near the main doors. (BAe plc)

In December 1940, 504 Squadron was replaced at Filton by 501, it in turn leaving in April 1941 as 263 Squadron arrived with Westland Whirlwind twin-engined fighter-bombers. After four months 263 moved to Charmy Down, the last operational squadron to be based at Filton in the Second World War. Two new tarmac runways were laid down, and 44 Group arrived as an OAPU carrying out work on Blenheims, Beauforts and Beaufighters, before it became 2 APU. Filton was also home to 528 RCU (Blenheims) from June 1943 to May 1944, after which it was employed on ferrying and flight testing duties.

After agreement was reached to build the massive Bristol Brabazon airliner, work commenced in March 1946 on a specially widened runway 8,150 ft long. Extensions were made to the main factories and a Brabazon assembly hall built, while Filton airfield was later transferred to the MoS. An RAF presence still existed, with Bristol UAS, 501 Squadron (AuxAF) and 12 RFS with Tiger Moths and Avro Ansons. By 1958 cut-backs had resulted in the disbandment of 12 RFS and 501 Squadron, but Bristol UAS was joined by 3 AEF and both soldiered on with their de Havilland Chipmunks. At the time of writing Filton is a BAe airfield, also used by Rolls-Royce (Bristol) for test flying. The Bristol UAS Chipmunks later gave way to SAL Bulldogs.

Meanwhile back in 1920 the Cosmos aero-engine company had collapsed, and the Air Ministry was anxious for Roy Fedden (later Sir Roy), designer of the very promising Cosmos 9-cylinder Jupiter radial engine, to develop his work. Bristol was approached with a view to purchasing the aero-engine assets of Cosmos, which it did for £15,000, acquiring five Jupiters, parts, drawings, patterns, tools and the services of Roy Fedden and a team of some thirty engineers. In September 1921 Fedden's Jupiter II passed a severe Air Ministry test, achieving 400 hp at 1,625 rpm. After exhibition at the Paris Air Show, licenced production was undertaken by the French Gnome-Rhone engine company, and eighty-one Jupiters were ordered from Filton by the Air Ministry for RAF use. Thereafter Bristol's aero-engine department was responsible for producing some of Britain's finest radial engines. As Bristol Aero-Engines Ltd it amalgamated with Armstrong Siddeley to form Bristol Siddeley Engines Ltd, which in turn became part of the Rolls-Royce Aero-Engine Division.

On 16 September 1932 a Vickers Vespa (rear) designed by Rex Pierson (right), with Roy Fedden's (left) Bristol Pegasus S engine, attained a world altitude record flown by Bristol chief test pilot C.F. Uwins (centre). The group was photographed after the event. (Rolls-Royce plc)

Bristol air-cooled radial engines under construction in the company's Aero-Engine Division works at Filton, late 1930s. (BAe plc)

Beauforts and Beaufighters being produced at Bristol's Filton factory in March 1941. (BAe plc)

Bristol's mighty Brabazon outside its assembly hall, Filton, c. 1949, prior to the application of its markings. (BAe plc)

Bristol's Armament Division was also a self-contained unit at Filton. Its main function was the design and production of power-operated gun turrets, an early example being fitted to Blenheim bombers. Outstanding was the Bristol B17 turret, with all-electric drive and housing two 20 mm Mk V Hispano cannon. Installed in Avro Lincoln four-engined heavy bombers, these were introduced too late for operational use in the Second World War.

After the war Bristol was heavily involved with design and production of the Type 170 Freighter, the giant Brabazon one-off airliner, Sycamore and Belvedere helicopters, the Britannia airliner and development of the Bloodhound SAM. By the 1960s changes within the structure of Britain's aircraft industry resulted in Bristol Aircraft Ltd (on separating from the aero-engine division it built airframes only), becoming part of BAC. As the Filton Division of BAC, Bristol's name was erased, Filton Division itself later being swallowed up by the BAe conglomerate.

BRISTOL BOXKITE AND EARLY BIRDS

The Bristol Boxkite, which first flew on 30 July 1910 at Larkhill, was based on a design of the French Farman brothers. The name was most appropriate and, as a two-seater machine, it was very suitable as a trainer. Typical of its day, all 'stick and string', the Boxkite featured biplane wings and tail unit, and was composed of steel tubing, wooden members, extensive wire bracing and fabric-covered surfaces. A 50 hp Gnome rotary engine provided the power and produced a top speed of 40 mph.

The first two Boxkites went to Brooklands and Larkhill Flying Schools, a third later going to Bristol's own new flying school near Stonehenge. The type was demonstrated over Salisbury Plain by Capt Bertram Dickson, and it became the first aircraft used on British Army manoeuvres, with Capt Dickson flying his Boxkite on a sortie over Salisbury Plain on 21 September 1910.

Boxkites were also sold abroad: two to Australia, three to India, eight to Russia, two to Belgian pilot Joseph Christiaens for use in Malaya and South Africa, others going to France, Bulgaria and Spain. The British War Office also ordered a number of Boxkites for RFC and RNAS service and the type became Britain's first production-line aeroplane. With two a week being turned out at Filton over eighty Boxkites had been produced by the end of 1912.

Other early designs from Filton at the time of the Boxkite included both biplanes and monoplanes. The first was the two-seat Bristol Glider, a biplane with front and rear elevators, which flew initially in December 1910. It was the intention to fit an engine in this machine, but this idea was abandoned and the glider scrapped. A Boxkite development, the Type 'T', a single-seat biplane with enclosed nacelle for the pilot, had a Farman-style curved front boom framework, four wheels in tandem pairs and a 70 hp Gnome engine. Five Type 'T' biplanes were completed and one entered the 1911 *Daily Mail* Circuit of Britain Race.

The first standard tractor biplane designed in Britain came from Filton in 1911. This Racing Biplane, designed by Robert Grandseigne, had unequal span wings, single interplane struts and a 50 hp Gnome engine. This machine crashed on its first flight in April 1911, the same year that Bristol built two single-seat monoplanes with wing-warping control, also fitted with a Gnome engine. The first machine suffered landing gear damage; it was repaired at Filton and exhibited at the Aero Show, Olympia, during March 1911. The second monoplane went to Russia for display there the following month. Another Bristol one-off was a developed Type 'T' biplane, the Challenger-England, which flew at Larkhill from November 1911 until 9 May 1912, when it overturned on landing and was written off.

Better-known were the Prier monoplanes, designed by Pierre Prier after he joined the Filton team. First came two single-seat racers entered in the *Daily Mail* Circuit of Britain Race in July 1911, but

This replica of a Bristol Boxkite was built by F.G. Miles Engineering Ltd for the film Those Magnificent Men In Their Flying Machines. *It had a Rolls-Royce Continental engine, and later passed to the Shuttleworth Trust at Old Warden. (Courtesy of the* Gloucester Citizen*)*

neither could participate due to pre-race accidents. This design was developed as a single-seat advanced trainer with a 50 hp Gnome and, later, a 35 hp Anzani engine. Ten Bristol Prier single-seat monoplanes were built, several going to Larkhill Flying School (run by Bristol), one to Spain and two to Italy. A two-seat version first flew in September 1911, with a 50 hp Gnome, increased weight and greater wing-span. All the Prier two-seaters were produced at Filton, one going to the Paris Air Show, four to the Italian army and others to Larkhill Flying School, Spain and Italy. The British War Office ordered one (256) for the RFC which was delivered in February 1912.

A basic Prier monoplane was developed as the Prier-Dickson (due to Capt Dickson's association with Prier monoplanes), with lengthened fuselage and fixed tailplane. The prototype flew in July 1912, to be followed by ten production machines sold to Bulgaria, Germany, Turkey, the Larkhill School and one (261) to the RFC. Filton produced five biplanes during 1912 designed by E. Gordon England: his GE1 tractor biplane with a 50 hp Clerget engine; two GE2s with relocated fuselage in relation to the lower wing, modified wings, and a 100 hp Gnome rotary and 70 hp Daimler engines fitted respectively; Two GE3s, intended for Turkey with 80 hp Gnome engines, had to be abandoned because of problems with the wing spars.

When Prier left Bristol's, monoplane design was taken up by Henri Coanda, his first being an updated two-seat Prier-Dickson with altered wing spars, side-by-side seating and a 50 hp Gnome. Tandem and side-by-side seat versions were built for Italy and Romania, and for military use Coanda produced his Military Monoplane with an 80 hp Gnome. Two were sold to the British War Office, until they were temporarily banned when one crashed on 10 September 1912, killing its RFC two-man crew. Attempts to sell the Military Monoplane to Romania and Italy failed, in the first instance due to a crash and secondly because the type was not up to Italian Army standard.

Coanda's biplane designs for Bristol started with the BR7 of which five were ordered by the Spanish Army. Seven were built, but performance was not up to expectations and Spain's order was cancelled. A single-float Coanda seaplane, evolved from the GE3, flew on 15 April 1913, but was forced to ditch in the sea with an overheated Gnome engine (fitted with a very close cowling) and was lost. The TB8 biplane, converted from Coanda's monoplane design, first flew in July 1913. It showed more promise, and also flew successfully as a twin-float TB8H seaplane. It became RNAS

Bristol Prier monoplane, powered by a 50 hp Le Rhone rotary, at Filton in 1911. (Author's collection)

Original Bristol TB8 biplane of 1913, with an 80 hp Gnome rotary engine. Note the landing gear. (Author's collection)

No. 15, and after another TB8 had been delivered to the RNAS a further twelve were converted as TB8 biplanes. One TB8 was shown in Paris during 1913, with bombsight and automatic revolving bomb carrier, and consequently licenced production of TB8s was undertaken by Breguet.

Six Coanda conversions went to Romania, and after the First World War erupted, the RNAS took delivery of thirty-six TB8 biplanes which had been modified with ailerons replacing wing-warping, twenty having a 50 hp Gnome engine, and sixteen a 60 hp Le Rhone. Before production ceased in February 1916, fifty-three TB8 biplanes had been completed including Coanda monoplane conversions.

Coanda and Bristol's experimental designer, Frank Barnwell (later chief designer), designed two more biplanes. The first one was the GB75 for Romania, powered by a 75 hp Mono-Gnome engine fitted with a large louvred spinner, which passed cooling air to the engine. Its first flight at Larkhill in April 1914 was vexing, the spinner being removed and several modifications made without success. The machine did not go to Romania, but with an 80 hp Gnome fitted it went to the RFC instead as No. 610. The second Coanda/Barnwell biplane design was a pusher with an 80 hp Gnome designated the PB8. Designed to supersede Boxkites only one machine was completed and that never flew.

Frank Barnwell was involved in an unusual design put forward by Lt C. Dennistoun Burney.

Capt Frank Barnwell, who, apart from an Australian break in 1922, was Bristol's chief designer for many years. (BAe plc)

The Burney Hydroplane had a watertight hull, small floats for support when at rest, hydrofoils, and engine-driven water propellers, which built up enough speed on water to raise the hull on the hydrofoils before power was transferred to the aircraft propeller. A special X department at Filton studied Burney's X1 twin-engine plans prior to Frank Barnwell's X2 emerging as a shoulder-wing monoplane with a single 80 hp Canton-Unne radial engine driving both air and water propellers via clutches. Trials revealed several problems and a new X3, with back-to-back contra-rotating water propellers, was built. This machine had an 80 hp Gnome, replaced quickly by a 200 hp Canton-Unne radial. Test pilot Harry Busteed started X3 trials in June 1914, but the machine crashed into a submerged sandbank and the project was abandoned.

SCOUT

Designed by Frank Barnwell, the Bristol Baby, or Scout as it became known, was a single-bay biplane with a 22 ft wing-span and an 80 hp Gnome rotary engine. Intended as a single-seat sporting biplane the prototype appeared during February 1914, and at Larkhill achieved a top speed of 95 mph. By May the Bristol Scout A had a revised cowling and an increase in wing-span to 24 ft 7 in. It was raced at Brooklands against the more powerful Sopwith Tabloid and, although losing by only a narrow margin, was flown by Harry Busteed at 97.5 mph. It was flown in the July Hendon to Paris Race by Lord Carbery, but was forced down in the Channel and sank; the pilot was picked up by a passing tramp steamer.

Two Scout Bs emerged from Bristol in August 1914, by which time Britain was at war. These aircraft had duplicated flying wires, enlarged rudder, wider landing gear track, revised aileron control cables, wing-tip skids moved to immediately below the interplane struts, and a modified engine cowling.

The first Scout B went to 3 Squadron armed with two rifles, one each side of the fuselage mounted at a 45° angle to avoid hitting the propeller. The second similarly armed machine went to 5 Squadron, and twelve Bristol Scout Cs (1602–1613) were ordered for RFC service on 5 November 1914, with a contract for twenty-four more machines (1243–1266) for the RNAS following on 7 December. These aircraft differed little from the B, external stiffeners on the cowling being deleted and the oil tank relocated aft of the cockpit. Delivery began in April 1915 and another order was placed with Bristol for a further seventy-five Scout Cs (4662–4699 and 5291–5327). A scarcity of Gnome engines in the summer of 1915 led to Le Rhone engines being installed in new RFC Scouts, and Gnomes in RNAS machines. These new RFC Bristol Scouts had the oil tank moved ahead of the cockpit, and some were fitted with an 80 hp Clerget rotary engine.

The updated Scout D had revised tail surfaces, shorter ailerons, streamline Rafwires on interplane bracing, increased dihedral, and wing-tip skids moved outboard. Eighty production Scout Ds went to the RFC with 80 hp Le Rhone engines, but a final version was powered by a 100 hp Gnome Monosoupape. Trouble with vibration and leakage in this engine resulted in the last twenty Scout Ds for RNAS service having 80 hp Gnome engines.

Bristol Scouts never fully equipped a RFC unit; instead one or two were allocated to each squadron in order to protect its two-seat reconnaissance aircraft. Thus Bristol Scouts operated over the Western Front with many RFC units. Later a number of them served at home as trainers, while others flew with Middle East units such as 30, 63, 67 (Australian) and 111 Squadrons. The last Bristol Scouts to fly in Mesopotamia were those of 63 Squadron's fighter flight at Samarra in November 1917.

In 1917 Bristol produced their more advanced Scout F, with a liquid-cooled 200 hp 8-cylinder Sunbeam Arab engine in a streamlined cowling and the radiator beneath. Performance from the first of the three Scout Fs (B3989) was good, with a top speed of 138 mph, but the unreliability of the Arab engine and its severe vibration was this aircraft's Achilles' heel. The last Scout F (B3991) was fitted with a new Cosmos Mercury radial engine to become the Scout F1, and test flights in April 1918 produced an excellent performance. However, the Mercury was shelved in favour of the Cosmos Jupiter radial, although testing of the Scout F1 continued after the Armistice.

This Bristol Scout 'D' with an 80 hp Gnome rotary engine, was destined for the RNAS. (Bristol)

Bristol Fighter

During 1916 Frank Barnwell designed a two-seat fighter biplane for Bristol in which the observer/gunner sat behind the pilot facing aft, the idea being to protect the pilot against surprise attacks from the rear. A prototype of this Bristol F2A Fighter (A3303) first flew in September 1916 powered by a 190 hp Rolls-Royce Falcon engine. A second machine (A3304) followed in October fitted with a 150 hp Hispano-Suiza, and an original order for fifty production aircraft was increased to 250. A two-bay biplane, the Bristol Fighter was armed with a single Vickers machine-gun firing forward through the propeller disc. The rear gunner was provided with single or twin Lewis guns on a Scarff ring in the rear cockpit; up to twelve 20 lb Cooper bombs could be carried beneath the wings.

Delivery of the F2As commenced in December 1916, and 48 Squadron RFC received its new aircraft in France during February 1917. Initially, heavy losses were suffered by F2A crews, but pilots soon learned to exploit the type by using it as a fixed-gun fighter with a gunner for rear protection. Bristol Fighters thereafter enhanced their reputation and became one of the more superior types of Allied aircraft in the First World War.

By April 1917 the improved F2B, with a 275 hp Rolls-Royce Falcon III, or 215 hp Sunbeam Arab, was reaching RFC squadrons, its increasing success warranting full quantity production by July 1917. Indeed F2Bs were scheduled to equip all Western Front fighter/reconnaissance units, and to satisfy the demand some production went to subcontractors; one such was GAC at Sunningend works, Cheltenham, which was awarded three contracts. Some GAC-built F2Bs were powered by Siddeley Puma engines.

When the RAF formed on 1 April 1918, Bristol Fighters were widely in use, and 22 Squadron's machines carried out the first sortie made by the new service at dawn that day. After the war F2Bs became standard Army Cooperation aircraft with RAF units, many being refurbished and 378 newly built. Some were converted as dual-control trainers, and post-war variants included the Mk II with tropical radiator, desert wheels and increased weight; Mk III

Bristol F2B two-seat fighter (E2448), one of many built at Filton during the First World War with a Rolls-Royce Falcon engine. (Courtesy of the Gloucester Citizen*)*

with Mk II features and an oleo tailskid; and the Mk IV with Handley Page wing slots, larger rudder with horn-balance, cambered fin and a strengthened landing gear. Two 112 lb bombs could be attached to underwing racks.

Bristol Fighters served in Ireland and Germany until 1922, 12 Squadron being last to leave the Rhine in July that year. In 1926 the RAF still had twelve squadrons of F2Bs in service, the last unit overseas to exchange its Bristol Fighters for Fairey Gordons being 6 Squadron in Iraq during 1932. Bristol F2Bs were also built for Belgium, Spain and Mexico. A Type 29 civil passenger-carrying version named the Tourer was produced and sold abroad to private customers.

A BRISTOL MISCELLANY (ONE)

At the time of its Scout development, Bristol built a one-off SSA biplane for France, the forward fuselage being a monocoque of sheet steel. With an 80 hp Clerget engine the SSA first flew in May 1914, but after landing gear damage this aircraft was delivered to France in crates. Another venture was the S2A side-by-side seater employing Scout wings and tail unit, but with a wider fuselage to accommodate its two occupants. Designed as a fighter for RNAS use, the S2A prototype (7836) first flew in May 1916, but was not accepted, the two built going to the RFC as trainers.

In May 1916 flight trials took place of the Bristol Twin Tractor, a large three-bay biplane fighter powered by two 120 hp Beardmore engines strut-mounted between the wings. A gunner sat in the nose, with the pilot further aft armed with a machine-gun on a flexible mounting. The initial flight of the first Twin Tractor (7750) at Filton in April 1916 proved the machine was underpowered and only two were built. At the same time Frank Barnwell, aware of diminishing timber stocks, designed jointly with W.T. Reid the metal MR1 biplane (A5177) with a monocoque Duralumin fuselage. Metal wings were intended but the Steel Wing Co. of Gloucester encountered snags in production, and it was not until late 1918 that the second prototype (A5178) flew as an all-metal machine. Power for the first MR1 was a 140 hp Hispano-Suiza engine, the second having a 180 hp Wolseley Viper installed. After A5178 crashed at Farnborough in April 1919, further development of the type ceased.

Work started on a Bristol monoplane single-seat fighting scout in 1915, when Frank Barnwell produced a shoulder-wing design of which two prototypes, the M1A and M1B, emerged powered by a 110 hp Clerget rotary engine. The definitive M1C was fitted with a 110 hp Le Rhone engine, and had a synchronized Vickers machine-gun mounted above the nose. Of 125 M1Cs built, some operated in the Middle East as ground strafers against the Turks, others flew as trainers, and six went to Chile in 1917 (one of these making the first crossing of the Andes by air in December 1918). After the war four became civil machines, one (G-EAVP), designated M1D as test-bed for the Bristol Lucifer engine, won the 1922 Aerial Derby flown by Larry Carter at 107.85 mph. However, on 23 June 1923 it crashed near Chertsey while competing in the Grosvenor Cup Race, and pilot Major Leslie Foote was killed.

Bristol's Type 23 Badger of 1919, a two-seat fighter intended as a replacement for the F2B, suffered ABC Dragonfly engine problems. The first Badger (F3495) crashed during its initial flight on 4 February 1919. In May a second Badger (F3496) flew powered by a 450 hp Cosmos Jupiter, and later a Mk II Badger (J6492) flew as a Jupiter and engine cowling test-bed. A four-engined triplane heavy bomber, the Bristol Braemar (C4296), was designed to bomb targets in Germany and made its first flight at Filton on 10 August 1918. With four 230 hp Siddeley Pumas in tandem pairs, initial tests were unsatisfactory, and a second Braemar (C4297) flew as a Mk II in February 1919 powered by four 400 hp American Liberty engines. This was an improvement, and a third aircraft (initially C4298) emerged to become the fourteen-seat civil

Bristol M1C monoplane fighter (C4910) with 110 hp Le Rhone rotary engine at Filton in 1917. (BAC, Filton)

Close-up of Bristol Jupiter engine installation in third Bristol Badger, J6492. Note the twin Vickers machine-guns fitted atop the front fuselage. (Rolls-Royce plc)

Pullman G-EASP. After flying in May 1920 this machine was entered as an exhibit at Olympia in 1921, but was later abandoned. A larger development was the Tramp, two of which were built (J6912 and J6913) with four 230 hp Puma engines mounted within the fuselage to drive two four-blade tractor propellers via a system of gears and shafts. Neither Tramp flew and both went to Farnborough where they were later scrapped.

Efforts to provide an affordable light private aircraft came from Bristol as the Babe of which three were built. The first, powered by a 35 hp Viale engine, flew in November 1919 piloted by

C.F. Uwins. The other two, each powered by a 60 hp Le Rhone, became Babe Mk IIIs G-EAQD and G-EASQ. Bristol's Babes never reached production status and all three were scrapped in 1924. Two Bristol one-offs were the Bullet and Seely, the former (G-EATS) being a Jupiter engine test-bed which flew in the 1920 Aerial Derby. It only managed 130 mph and modifications were carried out, but in the 1921 Aerial Derby the performance was still poor. Flown by Sqn Ldr 'Rollo' de Haga Haig in the 1922 Aerial Derby, the Bullet did come in second at 145 mph after having its specially designed large spinner removed! The Seely (G-EAUE), was in reality an updated version of the Bristol Tourer, with revised fuselage and tail surfaces, steel front fuselage members and a front skid to avoid nosing over. The engine was a 240 hp Siddeley Puma, but after civil trials at Martlesham the Seely was rejected. It transferred to Farnborough as J7004 with a Jupiter III incorporating an exhaust-driven supercharger, and was used experimentally.

A contender for a suitable commercial aircraft to serve with British airlines in 1920 was the Bristol Type 62 Ten-seater. A two-man crew sat in an open cockpit atop the fuselage, while the main cabin accommodated eight passengers. Intended for a Jupiter engine installation, the Type 62 was in fact powered by a 450 hp Napier Lion due to delays with the Jupiter. The first Ten-seater (G-EAWY) flew in June 1921 and was later used on Handley Page Transport's London to Paris route. This company was absorbed by Imperial Airways in 1924, which ended the Ten-seater's passenger work with that company due to its multi-engine-only policy for passenger aircraft. A second Ten-seater (G-EBEV) was Jupiter-powered and flew as the Express Freight Carrier Type 75A. The third and final Ten-seater was purchased by the Air Ministry as J6997 and became the Brandon flying ambulance, but was little used by the RAF after 1925.

Bristol came up with a two-for-the-price-of-one idea in the Bullfinch, which was operable either as a single-seat fighter, or two-seat fighter-reconnaissance type. In this the single-seat Type 52 was a high-wing monoplane, convertible to the Type 53 two-seat sesquiplane. Three Bullfinches were built (J6901–J6903), the first two as monoplanes, the third as a sesquiplane.

Bristol Bullet G-EATS of 1920, which tested and raced the Jupiter engine, in this case a Cosmos Jupiter I. (MAP)

Bristol Type 62 Ten-seater (G-EAWY), which flew with Handley Page Transport Ltd, c. 1922. (BAC, Filton)

Installation of 100 hp Bristol Lucifer radial engine in a Bristol Type 73 Taxiplane. (Rolls-Royce plc)

An unusual feature of the Bullfinch was its twin rudders beneath the rear fuselage. After official trials with both Bullfinch variants no production order was placed.

Meanwhile, Roy Fedden at Bristol's Aero-Engine Division was confident, after Gloster's Bamel set a new British air speed record of over 196 mph in December 1921, that a Jupiter-engined machine could better it by at least 20 mph. Hence the Bristol Type 72 monoplane racer (G-EBDR), which first flew in July 1922. This tubby machine incorporated a monocoque fuselage, totally enclosed engine with ducted cooling, manually operated retractable landing

Powered by a 690 hp Rolls-Royce Condor III engine, this was the Bristol Berkeley prototype, J7403, of 1923. Designed as a multi-purpose bomber it was the first of three built. (Author's collection)

Virtually a flying fuel tank, Bristol's Type 109 long-range biplane (G-EBZK) of 1928 with Jupiter VIII engine. (MAP)

gear, and wing root fillets. Several tests were carried out, but wing flexibility and aileron control problems arose, and the landing speed was too high. Further development costs were ruled out and the racer was abandoned.

In an effort to produce an efficient, economical commercial aircraft in the early 1920s, Bristol produced the Type 73 Taxiplane powered by a 100 hp Bristol Lucifer 3-cylinder radial. The prototype (G-EBEW) made its first flight in February 1923, accommodating a pilot and two passengers aft sat side-by-side in an open cockpit, with entry via a door in the port side. Since a C of A was unobtainable for the Taxiplane as a three-seater, Bristol changed the fuselage to tandem seating, thus creating the Type 83 Trainer. Of twenty-four

Type 83s built, several flew with Filton RFS, while others went to Bulgaria, Chile and Hungary.

Air Ministry Specification 26/23 of 1923 called for a large single-engined bomber to be powered by the 650 hp Rolls-Royce Condor III liquid-cooled engine. Among types submitted in the competition was the Bristol Berkeley, a three-bay biplane of which three were produced (J7403–J7405). As Bristol's first all-metal aircraft, its fabric-covered airframe was a combination of tubular steel and Duralumin. First flown by C.F. Uwins on 5 March 1925, the Berkeley had a top speed of 120 mph and did indeed fulfil all requirements of the specification, which included carrying a quarter-ton bomb load and a range of 1,200 miles. Even so no production Berkeleys were ordered, the three which were built ending their days on experimental work at Farnborough. During 1927, with a world record in mind, the Air Ministry sought an aircraft capable of flying more than 5,000 miles without refuelling. Response from Bristol was its Type 109 Long Range Biplane. A flying tanker, it carried almost 900 gallons of fuel in thirteen tanks of welded aluminium, and it was calculated that by cruising at 90 mph the biplane could achieve the record comfortably. Fitted with a 480 hp Jupiter VIII, the Type 109 (G-EBZK) made its initial flight on 7 September 1928, by which time the world record stood at 4,500 miles. This was considered too close to the extreme range of G-EBZK for it to make a record attempt and it was modified for Bert Hinkler to undertake a world flight. However, this had to be abandoned as, at over four tons gross, the Type 109 would be unable to use some of the aerodromes en route. Thus G-EBZK was used as a test-bed for the Jupiter XF engine before scrapping in 1931.

A BRISTOL MISCELLANY (TWO)

With new constructional methods developing at Bristol in 1925, a composite biplane with all-metal fuselage and swept-back wooden wings emerged at Filton as the Type 84 Bloodhound. This Jupiter-powered machine was first flight-tested on 8 June 1925 by test pilot C.F. Uwins. Metal wings were later fitted, the fin and rudder modified and the military-intended Bloodhound (G-EBGG) was used on Jupiter engine trials and as a test-bed for Hele-Shaw-Beacham propellers.

Meanwhile, for the Lympne Light Aeroplane Competition of the mid-1920s, Bristol produced the Brownie, a two-seat, low-wing monoplane with a 32 hp Bristol Cherub engine. Three were built (G-EBJK/L/M) weighing 500 lb empty, all with metal fuselages, the first having wooden wings and the other two metal. The trio were fabric-covered overall. The Brownie G-EBJK first flew on 6 August 1924, and in the Lympne Trials pilot C.F. Uwins won a second prize of £1,000 and £500 for best take-off and climb. The second Brownie (G-EBJL) suffered wing flutter and was withdrawn, the third machine (G-EBJM) making third place in the Grosvenor Trophy Race at 70.09 mph. This aircraft was converted to a single-seater with smaller wings, while G-EBJK became the Brownie Mk II with revised nose, rudder and steel wings of 17 ft 4½ in span. In this form it became Frank Barnwell's personal aircraft, until it hit a tree on take-off on 21 March 1928.

The experimental Bristol Laboratory Biplane flew from Filton in November 1925. It was built to investigate the streamline/drag effect of cowlings and front fuselage circular-sections in relation to engine overheating and cooling. With a Jupiter VI engine this aircraft carried out trials over a two-year period until written off at Filton in 1928 when the landing gear collapsed on touching down. In 1925 Bristol also had built a private venture two-seat Army Cooperation biplane, which was submitted as a replacement for the Bristol Fighter. Named Boarhound, one machine (G-EBLG) was built and first tested by C.F. Uwins on 8 June 1925. A two-bay biplane with a 425 hp Jupiter VI it was of all-metal fabric-covered construction, but was rejected in favour of the Armstrong Whitworth Atlas. Later a general purpose variant of the Boarhound

Built for the 1923/24 Light Aeroplane Trials, the Bristol Brownie (G-EBJK) with a 32 hp Bristol Cherub. (Bristol)

Bristol's chief test pilot for many years, Mr C.F. Uwins, who later became a Bristol company director. (BAe plc)

(G-EBQF) was built and first flew on 23 February 1927 as the Beaver. This was a private venture in a DH9A replacement competition, which was eventually won by Westland's Wapiti, and the two Bristol Beavers built were sold to the Mexican Air Force in 1928.

On 5 May 1926 a single-seat biplane racer made its initial flight from Filton piloted by C.F. Uwins. Named Badminton this machine (G-EBMK) had a 510 hp Jupiter VI engine, was of equal span and comprised a fabric-covered wood and metal structure. It was rebuilt in 1927 with tapered wings, raised upper centre-section, single interplane struts and an uncowled 525 hp Jupiter VI. On 28 July 1927 the Badminton crashed on take-off from Filton instantly killing the pilot F.L. Barnard. Test pilot C.R.L. Shaw was more fortunate when he parachuted to safety from the Bristol Type 101 two-seat fighter he was testing. This aircraft (G-EBOW), a single-bay biplane of mixed construction, was another private venture and first flew piloted by C.F. Uwins on 8 August 1927. It was undergoing steep diving tests with a 485 hp Bristol Mercury II engine when it broke up in mid-air on 29 November 1929.

In an attempt to enter the light civil market, the Bristol Type 110A five-seat passenger biplane was a fabric-covered metal design by Frank Barnwell. The only one built (G-AAFG) appeared at the Olympia Aero Show in July 1929, its first flight taking place on 25 October with a 220 hp Bristol Titan. This engine was later changed to a 315 hp Bristol Neptune, but the aircraft crash-landed at Filton in February 1931 and was scrapped.

Air Ministry Specification G4/31 called for a new general purpose aircraft to replace ageing RAF F2Bs and DH9As, such a requirement being anticipated by Bristol which had produced its Type 118. An all-metal biplane with 595 hp Jupiter XF this had Warren girder interplane struts, two-seat tandem accommodation for pilot and observer/gunner, and two .303 in machine-guns, one firing forward through the propeller and another in the rear cockpit on a ring mounting. Bombs were carried on racks beneath the fuselage. The first flight of the 118 (R-3) was on 22 January 1931, but in order to comply with the G4/31 requirements Bristol produced the updated Type 120 with a 650 hp Bristol Pegasus I radial. This aircraft's redesigned fuselage now incorporated a rotating transparent gun turret from Bristol's Armament Division. Neither the Type 118 or 120 (R-6/ K3537) was accepted for production, and both flew as Mercury and Pegasus engine test-beds respectively.

In response to Specification 2/34, which required an aircraft capable of research into flight at very high altitudes (50,000ft or more), Bristol produced the Type 138A High Altitude Monoplane (K4879). This was an all-wooden affair powered by an 840 hp Pegasus PEVIS supercharged engine, and with a span of 66 ft it was the world's largest single-seat aircraft. Its first flight was on 11 May 1936, and the following September K4879 climbed to 49,967 ft to gain the world altitude record for Great Britain. On 30 June 1937, after the Italians had beaten the record with an altitude of 51,362 ft in May, the modified Bristol 138A regained the record for Britain with a climb to 53,937 ft in 2 hours 15 minutes.

When Britain's Air Ministry issued Specification F5/34 its affinity with biplanes ended, for it

The Bristol Type 118 general-purpose biplane, later K2873, which first flew in January 1931 with a Jupiter XFA engine and became a Pegasus test-bed. (Bristol)

Bristol Type 138A High Altitude monoplane (K4879) with Bristol Pegasus PEVIS engine, which first flew in May 1936. (MAP)

Built to Specification F5/34, this Bristol Type 146 of 1938 had eight wing-mounted guns, and an 835 hp Bristol Perseus sleeve-valve engine. (MAP)

now required a single-seat, multi-gun interceptor monoplane fighter powered by an air-cooled radial engine. Bristol's Type 146 response was a low-wing, all-metal monoplane with eight (initially four) wing-mounted .303 in machine-guns. Landing gear was fully retractable and a 'bubble'-type sliding cockpit canopy was fitted. The intended powerplant was an 835 hp Bristol Perseus sleeve-valve engine, but an 840 hp Mercury was eventually fitted. The first flight was on 11 February 1938, the fighter (K5119) undergoing its trials at the A&AEE. On returning to Filton, the Type 146 collided with a display at the Empire Air Day Celebrations and received damage which caused its career to be abandoned.

Built to Specification A39/34, Bristol's Type 148 (K6551), a two-seat Army Cooperation monoplane, was designed for close-support, reconnaissance and artillery-spotting duties. An all-metal low-wing monoplane, with retractable main landing gear and faired fixed tailwheel, the 148 had an 880 hp Mercury IX engine. It first flew on 15 October 1937, but was damaged on its flight trials, a second machine (K6552) flying in May 1938. Production status was not reached, and the Type 148 was used as a Bristol Taurus engine test-bed.

One of two Bristol Type 148s built as Army Cooperation aircraft, K6551 was the first to fly in October 1937 and had an 840 hp Bristol Mercury IX engine. (Bristol)

The last biplane built by Bristol, the Type 123 was powered by a Rolls-Royce Goshawk steam-cooled engine. (Bristol)

Two designs to the same Specification F7/30 from Bristol comprised biplane and monoplane layouts. The first design was the Type 123 single-seat biplane, with fixed faired-in landing gear and a 695 hp Rolls-Royce Goshawk III steam-cooled engine, which made its initial flight in June 1934. This aircraft suffered Goshawk engine problems and the wing structure proved weak. Trials revealed very poor handling qualities, and further development was abandoned in favour of the Type 133 (F7/30) monoplane. Powered by a 640 hp Mercury, the prototype Type 133 (R-10) was a low-wing, single-seat fighter of all-metal construction with stressed-skin covering, semi-retractable landing gear, enclosed cockpit, and four .303 in Vickers guns, two in the front fuselage and two wing-mounted. Its initial flight was in June 1934, with a top speed of

260 mph, but during spinning tests in March 1935 the Type 133 went out of control and was totally destroyed, the pilot escaping by parachute.

The twin-engined, ten-seat Bristol Type 143 transport (R-14/G-ADEK), powered by 500 hp Bristol Aquila sleeve-valve engines, was built at Filton in 1935. Its first flight was on 20 January 1936, and the aircraft was demonstrated at Whitchurch the following September. Development of the sleeve-valve Aquila was abandoned, however, as was the 143 which was scrapped in the Second World War.

BULLDOG

In 1927 Specification F9/26 was issued calling for an all-metal day and night fighter to replace RAF Gamecocks and Siskin IIIAs. From Frank Barnwell at Bristol came his private venture Type 105 Bulldog, an unregistered single-bay biplane which first flew on 17 May 1927. Stiff competition was faced from other manufacturers, but in the event Bristol's machine was favoured. It was robust, easy to handle, highly manoeuvrable and could be dived safely at 270 mph. Of all-metal construction the single-seat Bulldog was fabric-covered except for the aluminium sheet-covered front fuselage. Power was a 450 hp Jupiter VII radial, and armament comprised two .303 in Vickers guns synchronized to fire through the propeller, and provision for four 20 lb bombs on underwing racks.

To improve spin recovery the prototype's fin area and shape was altered, the fuselage lengthened and a second prototype (J9480) ordered as the Mk II. After further trials the Bulldog was found easiest to maintain and repair, as well as having a faster top speed than its main rival the Hawker Hawfinch. In consequence the Air Ministry ordered twenty-five Bulldog Mk IIs from Bristol for RAF service (J9567–J9591), the last machine flying temporarily on Mercury engine tests as G-AATR before passing to the RAF. Bristol's company demonstrator (G-AAHH) went to Japan, where it flew with a Nakajima licence-built Jupiter.

Bulldog IIs were exported to Latvia, USA (US Navy), Siam, Australia, Sweden, Chile and Estonia. A further batch of twenty-three (K1079–K1101) went to the RAF in 1930 for service

Bristol Bulldog IIA (K2159), 500 hp Bristol Jupiter, seen here with 19 Squadron, RAF, 1932. (Author's collection)

Pictured in 1928, prior to installation of a Mercury IIA engine, the uncovered airframe of the Bristol Bullpup fighter. Note the main fuel tanks contained in the upper wings. (BAe plc)

with 17 and 54 Squadrons, while another Bulldog (R-1/G-ABAC) was used as a Bristol demonstrator and test-bed for Mercury III and Jupiter VI engines respectively. Then Bulldog demonstrator G-ABBB flew with internal structural improvements, a Jupiter VIIF engine and a gross weight of 3,530 lb. This resulted in an order for ninety-two Mk IIA Bulldogs for RAF service, a figure which eventually increased to over 260 machines as equipment for ten fighter squadrons. Bulldogs were the most widely used single-seat fighter in RAF service until 1936, and for several years provided 70 per cent of Britain's fighter defences, the last giving way to Gloster Gladiators in 1937. Apart from RAF machines, eight went to Sweden and four to Denmark for service in their respective air forces. One RAF Bulldog (K2188), converted as a dual-control trainer, led to fifty-nine Bulldog TMs (Training Machines) being built with redesigned tail unit and a $3\frac{1}{2}°$ sweepback to the upper mainplanes. Bulldog TMs went to the CFS, RAFC (Cranwell) and to 1, 3, 4 and 5 FTSs

An improved Bulldog IIIA (R-5), with Mercury IVA engine and top speed of 208 mph, failed to beat Gloster's SS19B Gauntlet in a competition for a Bulldog IIA replacement. The Bulldog IIIA crashed in March 1933, a second machine (R-7) flying the following month to become the Bulldog IVA prototype. This flew as a contender in the F7/30 competition, won by the Gloster Gladiator, but seventeen Bulldog IVAs went to Finland. These had lower wing ailerons, Mercury IVS 2 engine, long chord cowling and revised tail unit. One remains in the Finnish Air Force Museum.

A Bulldog development was the Type 107 Bullpup, basically a scaled-down Bulldog designed to Specification F20/27. Only one was built (J9051), the type losing out to Hawker Aircraft's Fury with its Rolls-Royce Kestrel engine.

BY JUPITER

The large Cosmos industrial concern acquired Brazil Straker's factory in Bristol during 1918 and, as Cosmos Engineering, the new Bristol management favoured aero-engine production against that of omnibus engines and Straker Squire motor cars which had previously been built on the premises.

Cosmos retained designer Roy Fedden (later Sir Roy), an employee of Brazil Straker since 1907, whose aero-engine interest had persuaded Brazil Straker to refurbish and build new aircraft engines during the First World War. The Cosmos take-over encouraged Fedden to develop the air-cooled radial engine concept, and his design for a single-row type with nine large bore cylinders, each with four valves, was accepted by the Air Ministry as the Jupiter. However, the war ended before it could be tried out in a combat aircraft, but on 18 June 1919 a Cosmos Jupiter engine was airborne for the first time in a Bristol Badger Mk II (F3496). This engine was later installed in the Sopwith 107 biplane (G-EAKI) flown by Harry Hawker in the 1919 Schneider Trophy Contest.

When Cosmos faced a serious financial crisis, things looked bleak for Fedden and Cosmos employees. Fortunately at that time the Bristol Aeroplane Co. were forming an aero-engine division and acquired Cosmos assets immediately. This included the services of Roy Fedden, five Jupiter engines and most of the Cosmos staff.

Having already seen his earlier Jupiter licence-built by Gnome-Rhone of France, and Bristol receive an order for eighty-one Jupiters for RAF service, Fedden continued to improve the design through his Mk V, VI and VII. The latter produced 460 hp, and in 1928 reduction gearing was introduced, allowing 2,200 rpm with reduced vibration and the driving of a large diameter propeller with greater efficiency. The developed Jupiter IX was rated at 525 hp, while a geared and supercharged Jupiter X gave 530 hp at 16,000 ft. At this altitude the original Jupiter I had produced 220 hp!

Jupiters were flown in competition with Armstrong Siddeley's Jaguar radials, but much lower maintenance costs and quicker turn round persuaded Imperial Airways to choose the Jupiter for their expanding fleet of transport aircraft. These included the de Havilland DH 66 Hercules, Handley Page HP 42/45 Hannibal class, Short Kent flying boats and the Handley

Close-up of a 9-cylinder, air-cooled, Bristol Jupiter radial, designed by Roy Fedden. (BAe plc)

Page W9 Hampstead, an early tri-motor biplane carrying fourteen passengers and a crew of two. Military types with a Jupiter engine included the Boulton & Paul Sidestrand, Bristol Bulldog, Gloster Nighthawk/Mars VI and Gamecock, Handley Page Hinaidi and Clive, Hawker Woodcock, Parnall Plover, Short Rangoon flying boat and the Westland Wapiti. Fedden's famous Jupiter was built under seventeen foreign licences, and some 7,100 were produced.

BOMBAY

With the issue of Specification C26/31 the Air Ministry sought an aircraft capable of functioning either as a troop-carrier with accommodation for up to twenty-four fully armed men, an air ambulance carrying ten stretcher cases, a freighter with the capacity to convey heavy spares and components, or as a heavy bomber with a crew of four able to defend itself against air attack. A successful contender was the Bristol Type 130, one prototype (K3583) being ordered in March 1933.

Designer Frank Barnwell used a multi-spar cantilever wing structure, a method developed at Filton after aileron and wing twisting problems had occurred during 1927 on the Bristol Bagshot twin-engine fighter. Only a few flight tests had been made with the Bagshot (J7765) before its development programme was abandoned, but experience gained with this aircraft led to Air Ministry research into all-metal cantilever wings. This proved invaluable when Barnwell introduced his seven spar wing to Bristol's new bomber transport. The wing, built in three sections, had two 750 hp Bristol Pegasus IIM3 engines mounted on the centre-section and was incorporated into the upper fuselage to give high-wing configuration. The fuselage was an oval monocoque with stressed-skin metal covering, and included five portions: nose with manually-operated Bristol gun turret, flight deck, centre, aft and extreme rear with tail gunner's position. A braced monoplane-type tailplane carried twin fins and rudders and the main wheels, on long oleo legs attached to the engine mountings with supporting stays to the fuselage, were housed in streamline spats.

The initial flight of the Type 130 (K3583) was from Filton on 23 June 1935 with C.F. Uwins at the controls. Satisfactory manufacturer's trials followed, and K3583 went to the A&AEE, where most of its military tests were undertaken by Flt Lt Bill Pegg (later Bristol's chief test pilot). Several improvements were made later, including two uprated 1,010 hp Pegasus XXIIs, a Scarff gun ring mounted in a shaped cupola for the tail gunner, the port side freight loading door modified to allow handling of heavy and/or bulky cargo with a built-in gantry, and three mobile water storage tanks provided for.

In April 1937 the 130 was named Bombay, and a production order placed for eighty Mk Is (later reduced to fifty) to Specification 47/36. New Bristol hydraulically operated gun turrets were now installed, a bomb-aimer's panel fitted, external fuselage racks increased the bomb load to 2,000 lb, a Vickers 'K' gun mounted in nose and tail turrets, wheel spats deleted, rudders altered, and a D/F loop and radio masts fitted. A four-man crew flew in the pure bombing role, but for troop-carrying three sufficed.

Due to Filton's preoccupation with Blenheim development, Bombay production was switched to Short Bros at Belfast. The first production machine (L5808) flew in March 1939 with a silver finish, but from L5819 onwards Bombays wore the standard camouflage pattern. First deliveries went to 216 Squadron in Egypt, others going to 117 Squadron (Khartoum) and 271 Squadron (Doncaster). Prior to the collapse of France in 1940 UK-based Bombays flew supplies across the Channel to Allied forces, although most Bombays were operational in North Africa, Sicily and over Italy. They flew as night bombers, carried supplies and made numerous casualty evacuation flights as air ambulances. Bombays of 117 Squadron flew a regular supply run between West Africa and Egypt, later operating as a support squadron for Allied forces in the Western Desert.

Powered by Bristol Pegasus XXII engines, this was the first production Bristol Bombay (L5808) built in Belfast under contract by Short & Harland Ltd. (BAe plc)

At the same time 267 Squadron's Bombays flew a Western Desert mail service, carried VIPs, made regular Malta flights, supported special units operating far out in the desert, and assisted in the evacuation of Greece by flying out troops and equipment. Thousands of Allied casualties were evacuated by Bombays; the No. 1 Air Ambulance Unit of the RAAF operated such mercy flights, their Bombays carrying large red crosses on wings and fuselage.

Blenheim

On 25 June 1936 Bristol Blenheim prototype K7033 made its first flight from Filton powered by two 840 hp Bristol Mercury VIII radials. Developed from the civil Type 142, the 142M (named Blenheim in April 1936) resulted from Air Ministry enthusiasm in the commercial machine's potential as a bomber. The civil 142, built for Lord Rothemere, first flew in April 1935 and, after being named 'Britain First', was presented to the Air Council. By September 1935 an Air Ministry order for 150 Type 142M bombers to Specification 28/35 had been placed.

Designer Frank Barnwell altered wing location of the civil 142 from low to mid position, allowing four 250 lb bombs to be carried in the 142M's fuselage. Armament comprised a .303 in Browning machine-gun in the port wing and a .303 in Lewis gun in a partly retractable dorsal turret. Military requirements were met by a strengthened airframe, and the crew consisted of pilot, navigator/bomb-aimer and gunner. A further 434 Mk I Blenheims were ordered in July 1936 and within a year twenty-four Blenheims a month were being produced at Filton.

Blenheims first entered RAF service on 10 March 1937, when 114 Squadron, Wyton, received its new bombers. Three other squadrons followed suit, and during 1938 twelve Bomber Command units took delivery of Blenheim Is. The type was exported to Finland, Yugoslavia, Turkey and Romania. A Blenheim IF night fighter variant had a ventral gun pack containing four .303 in Brownings. Most of these, painted matt black, were issued to seven Fighter Command squadrons.

An attempt to improve the Blenheim I emerged with the Type 149 Bolingbroke, which had a 3 ft nose extension. This came in for criticism and was replaced by an orthodox stepped-down design, with scalloping of the port side nose resulting in the Mk IV 'Long Nose' Blenheim. Powered by uprated 920 hp Mercury XV engines this version entered production early in 1939, and was soon operational with 53, 90, 101, 113 and 114 Squadrons of RAF Bomber Command.

This all-black Blenheim IF (K7159/YX-N) was a night fighter variant, here serving with 54 OTU. (Author's collection)

At the outbreak of the Second World War twelve Blenheim IV squadrons and one fighter unit with Blenheim IVFs (ventral gun pack similar to IF) were operational. A Blenheim IV of 139 Squadron was the first RAF aircraft to enter German air space during a reconnaissance flight on 3 September 1939. Next day 107 and 110 Squadron's Blenheim IVs joined Wellington bombers in a raid on enemy shipping in the Schillig Roads. Six Blenheim IV units flew to France with the BEF, and by the time Germany launched its offensive on 10 May 1940 twenty-two Blenheim IV squadrons were ready. By then many Mk I Blenheims had been transferred to the Middle East for service with eight squadrons, while 203 Squadron (Aden) was equipped with Blenheim IFs. In the Far East some time later, Blenheim Is flew with 11, 34, 60 and 62 Squadrons.

Meanwhile, from Britain, Blenheim IVs of Bomber Command's No. 2 Group attacked enemy invasion ports on the Channel coast, later extending their operations to targets on the Belgian, Danish, Dutch and Norwegian coasts. In one three-month period Blenheims sank 300,000 tons of enemy shipping, and in the first thirty-four months of the war flew 11,332 sorties to drop 3,028 tons of bombs.

Iraq-based RAF Blenheim Is moved to Egypt when Italy entered the war, and some IF conversions flew as escorts and defended Suez. RAF Blenheim I bombers assisted Greek Air Force Blenheim IVs over Greece, and 30 and 203 Squadrons helped cover the British evacuation from Crete. Soon afterwards Blenheim IVs began arriving in the Middle East to replace RAF Mk Is.

At Filton Bristol developed the high-altitude Blenheim V with 830 hp Mercury engines, and the Bisley two-seat close-support with protective armour and a four-gun nose section. The high-altitude variant was dropped, but from the Bisley evolved Blenheims VA, VB, VC (dual-control trainer) and VD (tropical). An order for 940 Blenheim VDs was placed, this type taking part in Operation Torch (the 1942 North African landings). However, the VD was overweight, a top speed of only 240 mph leading to many losses. Among these were ten machines from 18 Squadron, shot down in a raid on Tunisia in December 1942 during which Wg Cdr H.G. Malcolm earned a posthumous VC for his courage.

Blenheim Is were active against the Japanese from 1941, and in 1943 were joined by Mk IVs.

Blenheim IVL, V6083/FV-B, of 13 OTU (originally with 86 Squadron as BX-Y). Note the aft-firing twin guns in the blister beneath the nose, c. 1943. (BAe plc)

Another Blenheim pilot awarded a posthumous VC was Sqn Ldr A.S.K. Scarf of 62 Squadron, who flew his Blenheim back to base despite mortal wounds received while attacking a Japanese target.

RCAF Blenheims (Bolingbroke IVs) were American-equipped; some had Pratt & Whitney Twin Wasp engines, one fitted with large twin Edo floats beneath the engine nacelles and centre-section to become the Bolingbroke III seaplane.

BEAUFORT/BEAUFIGHTER

Bristol's twin-engined Beaufort torpedo-bomber prototype (L4441), built in response to Specification 10/36, first flew on 15 October 1938. Production Beaufort Is differed from the prototype in having a flat bomb-aimer's window, relocated exhaust pipes, revised landing gear doors and twin machine-guns in place of a single one in the dorsal turret. Uprated 1,130 hp Bristol Taurus VI engines superseded the earlier Taurus II, and armament was further increased later with two beam guns and a rear-firing blister offset to starboard beneath the nose. After more than 1,000 Beaufort Is had been built, the Mk II followed powered by Pratt & Whitney Twin Wasp engines. Out of 415 produced, 250 emerged as trainers with deletion of the dorsal turret. Another 700 Beauforts were built in Australia for the RAAF, all with Twin Wasp engines. Fifteen machines went to Canada for service with 149 Squadron (RCAF) in 1941, and six Mk Is were delivered to the Turkish Air Force.

Beauforts entered service with 22 Squadron in November 1939, and the following April carried out Coastal Command's first mine-laying sortie. They operated over the North Sea, English Channel, Atlantic, Mediterranean and Middle East, with 217 Squadron (Malta) and 39 Squadron (Western Desert). During April 1941 Beauforts made heroic attacks on the German battleships *Scharnhorst* and *Gneisenau*, the latter receiving a direct hit from Flg Off K. Campbell's Beaufort of 22 Squadron. For his bravery this officer was awarded a posthumous VC, as sadly he and his crew perished in the intense enemy defensive fire.

This Bristol Beaufort (L4449) later served with 22 Squadron, RAF Coastal Command. (BAC, Filton)

In the meantime Bristol's Type 156 Beaufighter, a private venture adopted by the Air Ministry to Specification F17/39, covered four prototypes plus a first production order for 300 aircraft. Filton began the design late in 1938 and combined the wings, rear fuselage, empennage and landing gear of the Beaufort with a redesigned front fuselage. Thus production could be switched from one type to the other with minimum disruption. Armament comprised four 20 mm nose-mounted cannon and (after the first fifty) six .303 in guns in the wings. Two 1,650 hp Bristol Hercules III radials were fitted, although Bristol's were asked to examine Rolls-Royce Merlin and Griffon engines as an alternative.

With their speed, firepower and an A1 radar installation, Beaufighters became Britain's first really efficient night interceptor. Beaufighter IFs had A1 Mk IV radar and, during the night air blitz of 1940–41, scored a number of 'kills'. By the end of May 1941 some sixty enemy aircraft had been shot down, by which time demand for Beaufighters necessitated production at Weston-Super-Mare, Stockport, and Blythe Bridge. Some eighty were modified for tropical use in the autumn of 1940, with extra fuel tanks for range extension. Mk ICs operated from Malta with 272 Squadron, differing from IFs in having special navigation and radio equipment.

A Beaufighter IIF had first flown on 26 July 1940 powered by two 1,280 hp Rolls-Royce Merlin XX engines, the first production machine (R2270) flying on 22 March 1941. This variant operated with eight night fighter squadrons starting with 600 Squadron at Colerne in April 1941. The IIF introduced a 12° dihedral to the tailplane which became standard on all following versions. One IIF (T3032) tested a dorsal fin extension to remedy the Beaufighter's tendency to swing on take-off, and this was standardized on later Mk Xs. Another IIF (R2274) was fitted with a dorsal gun turret, housing four .303 in machine-guns and designated Mk V. In its Mk VI form with two 1,670 hp Bristol Hercules VI or XVI, the Beaufighter could carry a 1,650 lb or 2,127 lb torpedo beneath the fuselage. After a Mk VIC (EL329) had been armed with eight underwing rocket projectiles plus two 250 lb bombs, a Beaufighter Strike Wing was formed at North Coates comprising 143, 236 and 254 Squadrons. Torpedo-carrying machines, nicknamed 'Torbeau', made a first successful strike against German shipping on 18 April 1943. Beaufighter VIs were sent to the Far East, where they eventually equipped eight RAF squadrons. The Japanese Army respected Beaufighters, naming them 'Whispering Death' as they flew in low and quiet to wreak havoc and destruction among Japanese targets on the rivers and jungles of Burma.

The major production version, the Beaufighter X, with 1,750 hp Hercules XVIII engines, carried a torpedo, bombs, rockets or a combination of these. Some Mk Xs had ASV radar fitted into a 'thimble' nose. In March 1945 Beaufighters from 236 and 254 Squadrons sent five

Nicknamed 'Torbeau', this Beaufighter X features dihedral tailplane and carries a torpedo beneath the fuselage. (Bristol)

U-boats to the bottom in two days. After the war a number of Beaufighter Xs became TT10 target-tugs, which were delivered between 1948 and 1950. The last flight of a RAF Beaufighter was that of a TT10 (RD761) on 12 May 1960 from Seletar, Singapore.

BUCKINGHAM/BUCKMASTER/BRIGAND

Intended as a Blenheim replacement, the Bristol Buckingham was delayed by Centaurus engine problems and official uncertainty over requirements. When it was available in 1944, tactical use of the Buckingham had been rendered obsolete by de Havilland's Mosquito. Thus a conceived high-speed day bomber was relegated to transport duties, but not before its armed concept had been proved in prototype and early production form.

Prototype Buckingham (DX249) first flew on 4 February 1943 minus armament, but the second machine (DX255) was fully armed when its initial flight took place shortly afterwards. Both aircraft had tapered tailplanes and small fins, but the third and fourth prototypes (DX259/DX266) featured rectangular tailplanes and larger fins. Powered by Bristol Centaurus IVs the prototypes were followed by 119 production machines with 2,520 hp Centaurus VIIs or XIs (the original order for 400 was cancelled). Early Buckinghams, armed with four nose-mounted, two ventral and four dorsal turret-mounted .303 in machine-guns, were later converted as high-speed transports joining sixty-five other machines completed to the same configuration. As a fast courier aircraft Buckinghams carried four passengers and a crew of three, among its duties being special flights to and from Egypt and Malta.

With the cut-back in Buckingham production 100 of the first ordered were built as advanced trainers, with dual controls and other necessary alterations. This created the Buckmaster, a side-by-side seater for pupil and instructor, the prototype of which (TJ714) first flew on 27 October 1944. It was followed by TJ717 the second prototype, and the first production Buckmaster (RP122) left Filton in 1945. Powered by two 2,520 hp Centaurus VII engines, the Buckmaster was, on its introduction, the most powerful and fastest trainer to have served in the RAF, with a top speed of over 350 mph. It could carry an air signaller aft of the pilot, and blind-flying and instrument training was fully catered for. Buckmasters served until the mid-1950s.

The Brigand, the last of this Bristol trio, was intended as a Beaufighter replacement, but its role was switched to that of a light bomber for overseas RAF service. Bristol used Buckingham

Bristol Buckingham day bomber. Note the mid-upper and ventral gun positions on DX255. (Bristol)

Powered by two Bristol Centaurus 57 radials, this Bristol Brigand TF1 (RH742) was a light ground-attack aircraft. The type entered RAF service in 1949 and operated against Malayan terrorists from 1950 to 1953. (Bristol)

wings, tail unit and Centaurus engines for the Brigand, but designed a new fuselage of smaller cross-section, with revised raised canopy and dorsal turret deleted. Of four prototype Brigands ordered, the first (MX988) made its initial flight on 4 December 1944. At the start of production the first thirteen machines (RH742–RH754) emerged as TF1 torpedo-bombers, but did not reach first line units, going instead to Gosport and Thorney Island Development Units.

Between 1946 and 1950 a total of 142 Brigands were delivered to the RAF, with 84 Squadron receiving its aircraft early in 1949. They also flew with 8 Squadron in Aden, and in 1950 replaced the Beaufighters of 45 Squadron in Malaya. Until 1953 the latter unit's Brigands flew numerous attack missions with bombs and rockets against Malayan terrorists.

Sixteen Brigands were Met. 3 weather reconnaissance aircraft, while a number became specially equipped T4 and T5 radar navigation trainers to provide instruction for radar navigators on night fighting techniques. A radar scanner was nose-mounted and the rear cockpit could be darkened.

Brabazon

More than twice as long and much taller than two Avro Lincoln bombers placed end to end, or three semi-detached two-storey houses, Bristol's mighty Brabazon airliner made her maiden flight on Sunday 4 September 1949. It was a day of triumph for Bristol and British aviation in general, for at that time this huge aircraft was the world's largest civil landplane.

Among thousands watching from vantage points around Filton were 350 representatives of the world's press, and from the control tower Lord Brabazon of Tara, after whom the aircraft was named, waved and cheered with the rest as Bristol's chief test pilot A.J. 'Bill' Pegg lifted the new airliner clear of the runway after a run of some 500 yards. As the great silver machine (VX206/G-AGPW) climbed steadily east most spectators showed their surprise, for taxying trials, which had started the day before, were expected to last for several days with a number of 'hops' preceding the first circuit. But Bill Pegg said afterwards he decided to take off while taxying to the west of the 2,750-yard runway.

After a 27-minute flight which covered a wide area of Gloucestershire and South Wales, the Brabazon made its landing approach. One could feel the tension as this massive aircraft headed for the runway with nose down and flaps lowered, but any apprehension was unnecessary. The main wheels touched smoothly and precisely, no jar, no bounce, and within 600 yards Bill Pegg had brought her to a standstill by employing the braking power of Rotol's propellers in reverse pitch. He commented afterwards: 'It was a very comfortable ride. Everything went as expected'. On this initial flight Brabazon weighed 210,000 lb at take-off, with a fuel load of 4,000 gallons. Altitude flown was between 3,500 to 4,000 ft, with a cruising speed throughout of 160 mph, an approach speed of 115 mph and the landing speed of 88 mph.

With a span of 230 ft, length 177 ft and height of 50 ft, the Brabazon was a fully pressurized, high-altitude, long-range monoplane. It was intended as a trans-oceanic airliner for BOAC to enable the airline to fly a non-stop service from London to New York. The fuselage comprised forward lounge, dining saloon and lounge bar over wing, while a rear lounge would be equipped

Bristol's chief test pilot Bill Pegg peers from the cockpit of the Bristol Brabazon. Note the extensive riveting. (BAe plc)

In this January 1948 scene at Filton, the massive Bristol Brabazon is almost structurally complete in its specially built assembly hall. (Courtesy Barry Duddridge)

With accompanying Bristol IIA Wayfarer (G-AIME), the huge Bristol Brabazon (G-AGPW) on test over the West Country. (Courtesy Pete Hicks)

with a cine-projector and radio. Baggage and mail compartments were below decks, toilets and galley beneath the dining saloon and the crew's quarters and rest area segregated forward.

Power was provided by eight 2,500 hp Bristol Centaurus XX engines enclosed within the inner wing as four pairs coupled to a gearbox. Each drove four sets of 16 ft-diameter three-blade Rotol contra-rotating propellers, with self-contained, constant speed, feathering and pitch reversing mechanisms. Each drive was independent and engines were accessible in flight.

Wing outer sections (65 ft span each) were attached to the main centre-section and contained flexible fireproof fuel tanks with a total capacity of 13,500 gallons. The tailplane had a 75 ft span and vertical tail surfaces a height of 50 ft. Twin wheels were fitted to each hydraulically operated landing gear leg, the track between the main wheels measuring 55 ft.

Brabazon was the end result of a government committee which was set up in December 1942, under the chairmanship of Lord Brabazon, to consider Britain's post-war civil aviation needs. Bristol's design staff reviewed the project, preliminary investigations extending throughout 1943. By 1944 work had started on a half-scale wing and fuselage section to test structural strength with applied loads, a forward fuselage section for air pressure tests and a 21 ft-wide portion of the wing to try out the unique twin Bristol Centaurus engine installation. Drawings for the full-size aircraft were issued in April 1945 and construction of the prototype progressed over the next four years. In the meantime the controversial removal of an entire village (Charlton) and the closing of a dual carriageway to make way for a new runway to enable Brabazon to take off reached government level. In the event officialdom won the day, Charlton villagers were rehoused at Patchway and Brabazon's runway completed. But this and other delays meant that Brabazon was in fact some twenty-seven months late in making its first scheduled flight.

When work started on the Brabazon Mk II it was supposed to be powered by four 7,000 hp coupled Bristol Proteus turbo-props, but the Brabazon's wing design fell short of the necessary requirements for such a scheme. It was also discovered that the expected fatigue life of the aircraft was 5,000 hours. Thus this £3 million project was abandoned, and G-AGPW and the partly built Mk II were relegated to the scrap heap in October 1953.

FREIGHTER/WAYFARER

A very early post-Second World War development by Bristol was its Type 170 short-range utility transport. This design had actually been conceived towards the end of the war, its shape determined largely by British military needs, one of which was the ability to airlift an Army standard 3-ton truck.

Of high-wing monoplane configuration the Type 170's bulbous nose was fitted with clam-shell doors, a raised flight deck well above the cargo hold and/or cabin, a fixed landing gear, and power provided by two 1,675 hp Bristol Hercules sleeve-valve engines. As military transport requirements were greatly reduced, the MoS financed two prototype aircraft on condition that Bristol covered tooling costs and built two extra prototypes. The two MoS machines, designated Mk I Freighter, retained the nose-loading doors, but Bristol's two, known as the Mk II Wayfarer, had a solid nose, side entrance/loading door and a strengthened freight floor (optional). The Freighters were strictly cargo-only transports, but the Wayfarers were made available in several configurations, including an all-passenger version with thirty-two seats, galley and toilet.

Of the MoS machines (VR380/G-AGVP and VR382/G-AGUT), the first to fly was G-AGVP on 2 December 1945 with C.F. Uwins at the controls. Next up was Bristol's Wayfarer (G-AGVB) on 30 April 1946, with G-AGVC following in June. Last to fly was G-AGUT in September, by which time production machines were rolling out of Filton works. However, after the first prototype underwent trials at Boscombe Down as VR380, performance was not up to expectations and new outer wing panels were fitted giving a span of 108 ft to allow a gross take-off weight of 39,000 lb. This necessitated more powerful engines, but although Bristol built a 'new Type 170' (G-AIFF) as a Mk XI with extended wings, it did not reach production status as available engine power was still insufficient. When the 1,690 hp Hercules 672 engine appeared, however, G-AIFF became a prototype for the updated Freighter Mk 21.

Meanwhile G-AGVC had flown as a demonstrator in North and South America before returning to the UK and participating in the Berlin Air Lift. Later it emerged as an updated

On lease here to SABENA, Belgium, from Air Charter Ltd of Southend, is Bristol Freighter Mk 32, G-APAV, c. 1958. (Aviation Photo News)

military freighter with 1,880 hp Hercules 238s, double oil coolers, increased fuel capacity and strengthened freight floor. Gross weight was now 42,000 lb. Unfortunately, in May 1949 there were no survivors when G-AIFF crashed into the English Channel while on trials, and the Type 170 programme suffered another blow in March 1950 when G-AHJJ crashed at Cowbridge, Glamorgan, undergoing trials with one engine stopped. It was later discovered that climbing on one engine could have an adverse effect on the rudder causing damage to the fin, which in the case of G-AHJJ had broken away from the aircraft. Modifications preventing a reoccurrence of this problem were introduced immediately, and Mk 21s were cleared for service. A convertible variant, the Mk 21E, had cabin heating, soundproofing and thirty-two removable seats.

The Mk 31 was a successful new series with a gross weight of 44,000 lb, two 1,980 hp Hercules 734 engines and a dorsal fin. A convertible version was produced as the Mk 31E, and a military variant with provision for supply dropping was the Mk 31M. The best-known Type 170 was perhaps the Mk 32, its fuselage lengthened by 5 ft, empennage increased in area, and capacity to carry three cars and up to twenty-three passengers. This version, developed for Silver City Airways to operate their Channel Air Bridge, was known as the 'Superfreighter'; the first machine (G-AMWA) was delivered from Filton in March 1953. This aircraft was later converted into a 'Super Wayfarer' carrying sixty passengers on Silver City's London to Paris service. When Silver City merged with Air Charter in 1962 to form British United Air Ferries, a fleet of twenty-four Type 170s was in use, this number increasing to forty-one by 1970. An export order for fifteen Mk IA Freighters came from the Argentine Government and were delivered between 1946 and 1947. One Freighter designated Mk 31C (XJ470) flew with the A&AEE at Boscombe Down from 1955, but was later modified with fully furnished cabin and sold in September 1968. Altogether 214 Type 170 variants were produced.

SYCAMORE

Late in 1944 a helicopter department was formed at Filton by Bristol, and Austrian rotorcraft designer Raoul Hafner joined the new section. MoS interest in acquiring a small helicopter on a par with Sikorsky's S-51 resulted in Specification E20/45, to which Bristol responded with its Type 171. Two experimental prototypes were ordered (VL958 and VL963), the first of which made its initial flight on 27 July 1947 piloted by H.A. Marsh. Named Sycamore, this helicopter was powered by a 450 hp Pratt & Whitney Wasp Junior, and featured a cabin built of light alloy, a

Seen here at Staverton in the 1960s, this Bristol Type 171 Sycamore was originally Bristol's demonstrator G-ALSX. As G-48/1 it passed to Westland under a government plan to 'rationalize' Britain's aircraft industry. (Skyfame)

tailboom of stressed-skin attached to a central engine and gearbox mounting, and a rotorhead fitted with three wooden monocoque blades. Also Wasp Junior-powered was the second Sycamore which flew in February 1948, and at that year's SBAC Farnborough Show a third Sycamore prototype (VW905) appeared in the Static Park as a Mk 2 powered by a 550 hp Alvis Leonides radial. Its first flight was on 3 September 1949, but on the next take-off the rotor disintegrated. After a strengthened rotor had been fitted development flying continued as production started on fifteen Mk 3s. These had a widened cabin to accommodate three passengers on the rear seat (prototypes were four-seaters), shorter nose and accessory drive transferred from engine to rotor gearbox (this maintained essential systems in the event of engine failure). The initial production batch included one HC10 and four HC11 ambulance and communication machines for Army Air Corps evaluation, four HR12s for RAF Coastal Command Rescue operations, and two Mk 3As built for British European Airways with a freight bay aft of the engine.

Main production Sycamore was the Mk 4, which had a taller landing gear, four cabin doors and the pilot's seat moved from port to starboard. Orders for Mk 4s included three HR50 and seven HC51s for the RAN, three Mk 14 for use of the Belgian Air Force in the Congo, fifty Mk 52s for the West German Army and Navy (as it was then), two HR13s and some eighty HR14s for the RAF's Air Sea Rescue Service, with 275 Squadron receiving its first Sycamore on 13 April 1953. RAF overseas units flying Sycamores included 194 Squadron (Malaya), 284 Squadron (Cyprus), and 110 Squadron (Borneo).

BRITANNIA

Bristol's response to MoS Specification 2/47 for a BOAC medium-range airliner was the Type 175 Britannia which, with a 130 ft wing-span, 103,300 lb gross weight (to include forty-eight passengers, luggage and 3,370 lb of cargo), and an expected 310 mph if powered by four Bristol Centaurus 662 engines, came nearest to Ministry requirements. Alternative Bristol Proteus turbo-props and Napier Nomad Compound engines were considered, but these gave no performance guarantee. On 5 July 1948 three prototype Centaurus-powered Britannias were ordered, with a proviso that the second and third machines could be converted to Proteus turbo-props, the third being fitted out to full airline standard. Then BOAC, looking to its African and Far East routes, favoured Proteus engines, ordering twenty-five Britannias which could be converted to Proteus later. But Proteus was plagued with problems, that is until Stanley Hooker became Bristol's chief engineer (engines). He redesigned the Proteus in 1950 as a Mk 3 and on test during May 1952 this engine was a big improvement.

In the meantime Bristol had updated Britannia for trans-Atlantic service, with increased tankage, 130,000 lb payload (including eighty-three passengers), and the main landing gear featuring four-wheel bogies instead of the twin-wheel units planned. Structural testing of Britannia resulted in a gross take-off weight of 140,000 lb, and the design was considered outstanding with its highly pressurized 12 ft-diameter fuselage, superb wings of 2,055 sq ft area incorporating large double-slotted flaps, and Messier main twin-bogie four-wheel landing gear retracting backwards into the inboard engine nacelles.

The first flight of the Britannia prototype G-ALBO took place on 16 August 1952 from Filton with Bill Pegg at the controls. Designated Series 101 this aircraft required only a few modifications, and it appeared the economical Britannia, nicknamed 'Whispering Giant' because of its quietness, would be a world beater. But it was not to be. During a demonstration flight to Dutch KLM airline officials on 4 February 1954, the second prototype (G-ALRX) suffered a Proteus engine explosion causing a fire in the wing. After force landing on the Severn mudflats the aircraft was written off. Three months later Britannia prototype G-ALBO was returned to Filton after a flap drive failure, leaving Bristol without a flyable version of the type until first production Britannia (G-ANBA) flew the following September. This aircraft had to fly prototype-intended tests, and Bristol had no hope of meeting Britannia or Proteus engine delivery dates.

A projected all-freight Series 200 did not materialize, but from it evolved the mixed-traffic Series 250 and long-range passenger Series 300, both with a 10 ft 3 in longer fuselage. By May 1955 a Series 310 was flying with integral outer wing tanks and powered by four improved Proteus 755 turbo-props, the gross weight now being 185,000 lb which included 133 all-tourist class seating accommodation. Because of Proteus icing problems in 1956, BOAC did not fly Britannias until 1957, but Israel's El Al airline ordered three long-range Series 313s and flew them regularly between Tel Aviv and New York. BOAC operated eighteen Series 312s on its North Atlantic route, one Series 301 built (G-ANCA) crashed near Bristol with the loss of fifteen lives on 6 November 1957, and two Series 302s, minus long-range tanks, were sold to Aeronaves De Mexico. Some Britannias were built in Belfast by Short Bros, while new and second-hand machines were sold to airlines in Argentina, Canada, Cuba, Belgium, Burundi,

Prototype Bristol 175 Britannia 100, G-ALBO, at 1952 SBAC Show, Farnborough. (BAC, Filton)

Czechoslovakia, Ghana, Eire, Kenya, Liberia, Pakistan, Spain, Switzerland, Uganda, America, Zaire and several independent UK operators. After the Pan American Airways Boeing 707 jet airliner had crossed the Atlantic on 26 October 1958, it sounded the death knell for Britannia and other piston or turbo-prop airliners serving major long-distance routes.

Britannias operated successfully with the RAF from June 1959 until January 1976, giving the service its first turbo-prop transport. Twenty Britannia C1s, developed from the Series 250 multi-role freighter, served with 99 and 511 Squadrons, and were joined by three Series 252 long-range troop-carriers known as C2s. The RAF's twenty-three Britannias operated long-range strategic missions world-wide. When Belize suffered from the effects of 'Hurricane Hattie' during November 1961, Britannias from 99 and 511 Squadrons flew out medical teams and returned with evacuated families.

TYPE 173/BELVEDERE

Designated Bristol Type 173, Britain's first tandem-rotor helicopter was powered by two 575 hp Alvis Leonides piston engines driving two sets of Sycamore rotors and control systems. Drive was via a freewheel clutch and, with both rotor gearboxes interconnected by a shaft, in the event of one engine failing the other could drive both rotors.

Two prototypes were ordered to MoS Specification E4/47 calling for a twin-engined, ten-seat (or more) helicopter driven by a pair of Sycamore dynamic systems. The first machine (G-ALBN) emerged in 1951 with a slim semi-monocoque fuselage in silver, blue and white, three-blade rotors, steep dihedral tailplane, fixed four-wheel landing gear and large passenger windows. Attempts at flight in 1951 were delayed by resonance problems, and the first hovering flight was made on 3 January 1952, with test pilot C.T.D. 'Socks' Hosegood at the controls. Further ground resonance trouble, a tendency for the machine to take off at a steep nose-high angle and fly only backwards meant further delays, and G-ALBN did not make its first proper flight from Filton until 24 August 1952.

After appearing at that year's SBAC Show, the Type 173 prototype underwent RAF evaluation as XF785, and during 1953 naval trials took place aboard HMS *Eagle*. This machine was subsequently fitted with four-blade rotors, and a straight tailplane with oblong fins at the tip. A second prototype (G-AMJI) intended for BEA had stub wings fitted fore and aft, with a modified landing gear incorporating castoring front wheels. Designated Type 173 Mk 2, this machine first flew on 31 August 1953. It was later transferred to the RAF for trials as a maritime reconnaissance helicopter, serial XH379, with stub wings removed and the upswept tailplane of the Mk 1 reinstated. It was leased to BEA in August 1956 for trials, but withdrawn from use after an accident at Filton in September.

Three more prototypes (Type 173 Mk 3) were ordered by the MoS with 850 hp Alvis Leonides, four-blade metal rotors in place of the earlier wooden type and a higher rear pylon. First of these Belvederes (XE286), built at Weston-Super-Mare, flew on 9 November 1956. The second and third machines (XE287–XE288) were not flown despite completion, XE288 having a shorter fuselage and longer stroke landing gear as the Type 191 naval variant. This secured an order in April 1956 for three Alvis Leonides Major-powered prototypes, and sixty-five production aircraft with Napier Gazelle turboshaft engines. However, Bristol's received a severe blow when the Admiralty cancelled its Type 191 order in favour of the licence-built Westland-Sikorsky S-58.

However, RAF requirements included a personnel/paratroop transport and casevac helicopter, with externally slung load capability. Bristol produced this as the Type 192, and an order for twenty-six Gazelle-powered machines was placed during 1956. After Gazelle engine

Bristol/Westland Type 173 Mk 2 Belvedere with V-tail at the 1954 SBAC Show. This machine (XH379) was originally registered to Ministry of Supply as G-AMJI. (MAP)

tests had finally proved satisfactory, the first prototype Type 192 made its initial flight on 5 July 1958 at Weston-Super-Mare, to be joined later by nine pre-production aircraft. These originally had wooden rotor blades, anhedral-braced tailplanes and end-plate fins. All were raised to full production standard with metal rotor blades, compound anhedral tailplanes, powered flying controls, sliding doors, improved air intakes and larger low-pressure wheels.

Named Belvedere HC1, the first three pre-production machines (XG453/454/456) formed as a Trials Unit at Odiham in October 1960, and on 15 September 1961, 66 Squadron was first to be equipped with Belvederes at Odiham (later moved to the Far East). By then Bristol's helicopter division had been taken over by Westland Aircraft Ltd, which continued Belvedere production until June 1962 and retained product support while the type remained in service.

Belvederes served with 72 Squadron from November 1961 until September 1965, when they were replaced by Westland Wessex. In Aden 26 Squadron conveyed commandos from the aircraft carrier HMS *Centaur* into Tanganyika with its Belvederes during the 1963 rebellion, and later supported the Army in South Arabia during the Radfan operations. From Labuan, Borneo, 66 Squadron flew its Belvederes throughout the 1962–6 Brunei campaign. With the disbandment of 66 Squadron at Seletar in March 1969, Belvederes were retired from RAF service.

TYPE 188

A design and development contract to Specification ER134T was issued in 1953 calling for an experimental aircraft capable of sustained speeds of Mach 2 plus. It was required for kinetic heating research and the evaluation and proving of engines, structures and systems for hypersonic aircraft. Bristol was chosen from among a number of interested manufacturers and an initial contract signed for three Type 188s, one as a static test-bed, the other two for flight testing (XF923 and XF926). The design involved a long pencil-like fuselage, originally to be fitted with thin swept wings and two Rolls-Royce RA24R Avon turbojets located mid-wing. After wind tunnel tests with scale models the wing planform was altered to have rectangular inner sections, the outer portions having swept-back leading and straight trailing edges. An all-moving tailplane was fitted at the top of the fin, the main landing gear retracting inwards and the nosewheel, with its twin wheels, retracting into a bottom fuselage well. Construction of the Type 188 was to be from stainless steel, which raised many problems concerning fabrication and welding in this material.

Bristol Type 188 supersonic research aircraft (XF926) in landing configuration at Filton in 1963. (Peter R. March)

Following several types of engine trials, including Avon 200 series, de Havilland Gyron Junior and Rolls-Royce AJ65s, it was decided to fit two Gyron Junior DGJ10Rs producing 20,000 lb thrust at Mach 2 when at 36,000 ft altitude. This engine could be run at continuous supersonic speed, and introduced features like stainless steel and titanium into engine construction, as well as high temperature oil and fuel systems. Location and arrangement of these large turbojets facilitated alteration of engine air intake shapes and variable afterburner nozzles.

First Type 188 (unmarked) was delivered to RAE, Farnborough, in May 1960 for airframe testing, later moving to RAE, Bedford. The second machine (XF923) started taxying trials in April 1961, but due to numerous technical problems it was a full year before its first flight on 14 April 1962, when chief test pilot Godfrey Auty flew the machine from Filton to the A&AEE. The third prototype (XF926) made its initial flight on 29 April 1963, and on one test flight attained Mach 1.88 at 36,000 ft. But neither aircraft reached the speeds required of them, and the Type 188's career ended quicker than expected. Despite its extensive fuselage tankage, consumption was too great to allow endurance in the desired speed range. Both Type 188s were SOC in November 1966 and relegated as gunnery targets, but XF926 was retrieved and is now at the Cosford Aerospace Museum.

TYPE 221

On 10 March 1956 the Fairey FD2 Delta (WG774) established a new absolute world air speed record for Britain in the hands of Fairey's chief test pilot, Peter Twiss. It flew twice over a measured course at 38,000 ft attaining a speed of 1,132 mph. This well exceeded the previous record (set by the North American Sabre at 822 mph), and it was also the first record holder to have achieved over 1,000 mph.

A second FD2 was built (WG777) and both aircraft were used in a wide variety of research work at the RAE, Bedford. Then in 1959 Specification ER193D was issued for a high-speed research aircraft to be fitted with a completely new ogival wing as part of the Concorde supersonic airliner development programme. The FD2 was chosen and a contract issued to Fairey for one FD2 to be fitted with an ogival wing. This order, however, was passed on to Bristol via Hunting Aircraft, the first FD2 (WG774) transferring to Filton in September 1960 as the Bristol 221. When Bristol was absorbed into BAC, the aircraft became BAC 221, and conversion included a sharp-

The Bristol (later BAC) Type 221 ogival wing research aircraft WG774, now exhibited at the FAA Museum, Yeovilton. (Peter R. March)

edged ogival wing, longer fuselage, long-stroke landing gear legs, telemeters, updated avionics and uprated 14,000 lb thrust Rolls-Royce Avon RA 28R turbojets. Both FD2s had incorporated needle-noses that could be lowered to improve visibility when landing or taking off ('droop-snoots'), a feature to be incorporated in Concorde. The BAC 221 made its first flight, piloted by Godfrey Auty, from Filton on 1 May 1964. An extensive flight test programme followed until two years later when WG774 returned to Bedford for investigations into ogival wing high-speed characteristics. This aircraft is now an exhibit in the Fleet Air Arm Museum, Yeovilton.

CONCORDE

Over a quarter of a century ago the first Concorde prototype made its initial flight from Toulouse on 2 March 1969, yet this beautiful supersonic airliner is as popular as ever, continuing to operate Air France and British Airways transatlantic services. Origins of this Anglo-French design, which has a 93 per cent plus reliability record, began in the 1950s with a Ministry of Aviation contract issued to comply with a recommendation put forward by a Supersonic Transport Aircraft Committee. This comprised a Supersonic Transport (SST) of delta wing planform, with slender oval fuselage, airframe of titanium and steel, a range of 3,500 miles and a speed of 1,200 mph (Mach 1.8).

In response Bristol proposed a transatlantic airliner with accommodation for up to 130 passengers, powered by six Bristol Siddeley Olympus turbojets. However, Ministry officials asked for a smaller version seating 100 passengers and powered by four instead of six engines. Meanwhile France's Sud-Aviation were planning an SST 70/80 seater, smaller than Bristol's design but otherwise very similar. Britain needed a foreign partner in its SST project and consequently, after a meeting at Le Bourget in 1961, France agreed to finance an SST aircraft on a 50/50 basis, with BAC and Aerospatiale sharing airframe production and power provided by four Rolls-Royce Olympus 593 turbojets. Named Concorde in 1963, the new airliner was to have uprated Olympus 593Bs providing Mach 2 plus cruising speed at 51,300 ft. The Anglo-French arrangement resulted in assembly lines at Filton and Toulouse, with BAC producing nose, tail and engine installation, and Aerospatiale the wings, centre fuselage and landing gear.

The first British Concorde, G-BSST (002), flew initially from Filton to Fairford (Concorde's UK flight test base), on 9 April 1969 piloted by Brian Trubshaw. The first British pre-production Concorde, G-AXDN (01), emerged on 17 December 1971, and in France Aerospatiale's second pre-production machine (F-WTSA) made a fully automatic landing in January 1973. The following month it flew non-stop from Toulouse to Iceland and back with excess payload in a time of 2 hours 9 minutes at Mach 2.

On 28 July 1972 BOAC ordered five production Concordes and Air France four, with reservations on seventy-four more Concordes being placed with sixteen international airlines. However, with rapidly rising fuel costs caused by the 1974 world oil crisis, and the environmental protests made world-wide relating to Concorde's sonic boom and exceptional airport noise, the aircraft's future potential as a 'best-seller' was doomed. Even the US Government banned the Anglo-French aircraft from using American airports, although this was later lifted, and some circles in British and French aviation considered the American attitude as sour grapes because American SST plans had been dropped.

In the event no export order for Concorde was fulfilled and only sixteen production machines were built to fly with Air France and British Airways. The latter joined with Singapore Airlines in operating a London to Singapore service via Bahrein, while American operator Braniff leased Concordes from both British Airways and Air France in 1979 to fly subsonic services between Washington, Dallas and Fort Worth.

In the meantime Air France had started its Paris to New York service on 22 November 1977, and British Airways a London to New York route on 12 February 1978. Earlier, Air France began the world's first supersonic scheduled passenger service flying from Paris to Rio de Janeiro via Dakar and Caracas on 21 January 1976, later extending its Paris to New York route into Mexico City. During 1986 British Airways inaugurated a scheduled Concorde service to Miami, and on 2 December that year a chartered Air France Concorde landed in Paris, having carried ninety-four passengers around the world in eighteen days, a total flying time of 31 hours 51 minutes. At the time of writing Concorde is still the only SST in service anywhere, and during its twenty-five years and more of service has made a number of records, the most

Final assembly of BAC/Sud Aviation Concorde 002 (G-BSST) at Filton. Its first flight was on 9 April 1969. (Courtesy Mr & Mrs A.R. Hacker)

An excellent shot of the first Filton-built pre-production BAC/Sud Aviation Concorde (G-AXDN). This shows clearly the landing gear, engines and 'droop-snoot' nose on landing approach. (Courtesy Pete Hicks)

outstanding being that of G-BOAC, which crossed the Atlantic four times (London to Gander and back) in one day (1 September 1975) during route proving trials.

BLOODHOUND (SAM)

In a Government Defence White Paper of 1957 it was proposed to replace manned British fighter aircraft in the RAF with SAM-guided missiles. One of these important weapon systems was to be the Bristol/Ferranti Bloodhound produced at Filton. This SAM was among the most advanced of its kind in the world, and it was accepted on a scale never before achieved by a European system. In its Mk 1 and 2 form Bloodhound was ordered by Sweden and the Royal Air Force, while Australia ordered Mk 1s and Switzerland Mk 2s.

Basically Bloodhound operated on a semi-active homing radar guidance system in which a powerful ground radar picked up the target, 'locked' on to it, and illuminated the target aircraft with a narrow pencil beam of radar energy. The target reflected this energy and the missile was fired to home in on the reflection. The system comprised Target Illuminating Radar, several missiles and launchers, launch control post and the ancillary power-generating and handling equipment.

The Bristol Bloodhound Mk 1 was propelled by two Bristol-Siddeley Thor ramjets at a speed of Mach 2 after launch by four Gosling rocket-boost motors. From 1958 Bloodhound Mk 1s were operated in units of sixteen by Air Defence Missile Squadrons, which came under Fighter Command control. They were used mainly to defend RAF V-bomber bases or Douglas Thor surface-to-surface ballistic missile sites. Bloodhound 1, with Ferranti control post and radar guidance, although having a range of 40 miles was vulnerable to jamming of its pulse radar and had poor low-level capability. Consequently Bloodhound 2 superseded the Mk 1 into RAF service.

Bloodhound 2 can be transported by air, and due to its use of continuous wave rather than pulse radar is more resistant to counter-measures. Fuel capacity and range increased (50-mile

Bristol Bloodhound Mk II SAMs of 25 Squadron's 'A' Flight guard RAF Bruggen in Cold War days. (Peter R. March)

radius in published data), target interception altitudes can alternate between 59,000 ft down to well under 1,000 ft, and once detonated by its proximity fuse the warhead explodes to send a lethal concoction of metal particles spinning through the air in a circumference of some 120 ft. The Bloodhound was in service continuously with the RAF from 1957 until 1991.

CONTRACTS

During its long history Bristol also produced a number of other manufacturers' aircraft under subcontract, including many BE 2 series in the First World War, Parnall Panthers and Armstrong Whitworth Siskin IIIAs in the 1920s, Hawker Audaxes in the 1930s and Tempest IIs during the Second World War.

CHAPTER 4
Parnall's of Yate

George Parnall & Co., normally associated with Yate, Gloucestershire, was initially established at the Coliseum works, Park Row, Bristol. These premises were acquired by the Avery-owned firm of Parnall & Sons Ltd during the First World War for aircraft production. Under Admiralty contracts the company built Avro 504s, Sopwith (Fairey) Hambles, Short Seaplanes and Short Bombers. In addition Parnall produced its own Scout (Zepp Chaser) and two-seat Panther naval spotter reconnaissance biplane.

After the war, George Parnall, managing director of Parnall & Sons, disagreed with Avery's over its future business ideas in shopfitting and its lack of support for his aviation plans to further company interests. He therefore resigned from the Avery Group, leased the Coliseum and formed his own company. As well as manufacturing shopfronts and fittings for department stores, the necessary equipment for constructing aircraft was quickly set up alongside the cabinet makers.

George Parnall offered the post of chief aircraft designer to Harold Bolas, who had produced the Panther for Parnall & Sons earlier. Bolas immediately accepted, and his Puffin naval amphibian emerged, followed two years later by the Plover naval fighter. With the exception of the Puffin, which flew from the Isle of Grain Naval Air Station, Parnall aircraft were test-flown from Filton under an arrangement with Bristol until George Parnall & Co. moved to Yate.

A rare shot of Harold Bolas, chief designer to George Parnall & Co. at Bristol and Yate until 1929. He is standing before Parnall Elf G-AAFH. (Courtesy Norman Hall-Warren)

With the expansion of his business, George Parnall acquired three more local factory sites and offices in Bristol and London. However, after receiving Air Ministry contracts to refurbish and newly build DH9As for the RAF, George Parnall acquired Yate aerodrome, Gloucestershire (now part of Avon), 9 miles from Bristol. During the First World War the government had established a maintenance base at Yate known as Number 3 (Western) Aircraft Repair Depot. Facilities included brick-built engine repair shops, large wood and asbestos flight sheds and an aerodrome. As far as George Parnall was concerned it was tailor-made for his purposes; his company was now able to design, construct and test-fly its aircraft from a single site. The main office block and works were at the west end of the aerodrome, while the design offices were located at the opposite end of the airfield in the old RFC/RAF headquarters building. But take-off runs were limited to 1,500 ft, albeit into a prevailing wind, and pilots had to climb over works, hangars, main railway line, rail sidings and a group of houses.

During his time with George Parnall Harold Bolas produced a varying mixture of military and civil types, with names like Pixie, Perch, Pike, Peto, Pipit, Prawn, Parasol, Imp and Elf. He seriously studied a 'bodyless' monoplane design whereby a flying-wing, incorporated into a very wide fuselage, produced a continuous aerofoil section for both the wing and body. An excellent theory, although as a model this design proved disappointing in wind tunnel tests and was abandoned. Bolas left Parnall's and the UK for America in 1929, where he became a partner in the Crouch-Bolas Aircraft Company.

A majority of Parnall-designed aircraft were prototypes or used experimentally, and by 1932 George Parnall was selling off some assets; top works at Yate went to Newman's electrical engineers, and the Bristol-based shopfitting business to his old rival Avery's. By 1935 the RAF expansion scheme was under way, and with one already obsolete Parnall aircraft nearing

Aerial view of the Parnall works and airfield at Yate. Note the close proximity of the main railway line and sidings at the rear of the premises. (T.I. Jackson Ltd, Yate)

George Geach Parnall, founder of George Parnall & Co., poses in front of the first Parnall Elf biplane. (Parnall & Sons)

Close-up of Frazer-Nash 'lobster-back' segmented gun turret built by Parnall Aircraft Ltd, and fitted as here to Hawker Demon two-seat fighters, a number of which were converted at Yate to Turret Demons by Parnall's. (BAe plc)

In this aerial view of Parnall's Yate works, following the German air attack of February 1941, the severity of the damage caused is apparent. (T.I. Jackson Ltd, Yate)

Phoenix risen: Spitfire wing leading edges taking shape in the rebuilt Parnall Aircraft works at Yate in the latter years of the Second World War. (T.I. Jackson Ltd, Yate)

completion at Yate (G4/31 general purpose), George Parnall sold the remaining Yate site and retired. He continued living in Bristol, but on 23 May 1936 he suffered cerebral haemorrhage at his Clifton home and died on 21 June. He was buried in the parish churchyard at St Gennys in his native Cornwall.

The Yate site was acquired by Nash & Thompson Ltd (later Parnall Aircraft Ltd), noted for its power-operated aircraft gun turrets. Its 'lobster-back' type was fitted in the rear cockpit of Hawker Demon two-seat fighters, and as war in Europe seemed inevitable orders flowed in to Parnall Aircraft Ltd for Frazer-Nash turrets to equip Whitley and Wellington bombers, Blackburn Bothas and Short Sunderlands. Parnall Aircraft also undertook the manufacture of Spitfire airframes.

With the Second World War only five months old, Parnall's was targeted by a German Heinkel He 111 bomber on the afternoon of 27 February 1941. This attack on the Yate works cost the lives of over fifty employees, injured many others and caused considerable damage to the works. On 7 March a repeat raid was made, again by a lone Heinkel He 111, this time with less casualties, although material damage to the factory was more serious. Production ceased temporarily, but only eight days later started up again at Dursley, where a satellite works had been rapidly set up. Numerous other sites around Gloucestershire and Bristol were used by Parnall's during the war, but most impressive was the large modern factory rebuilt on the site of the old bombed-out Yate works. Parnall Aircraft was in full production there well before the end of the war, building nose and tail turrets for Lancaster bombers, Spitfire airframes, components for Lincoln bombers and, later, parts for Gloster Meteor jet fighters. This factory employed 3,500 people, but immediately after the war it switched from aircraft to domestic appliance production.

PANTHER/PUFFIN/PLOVER

Although not strictly associated with Yate, the first two machines here were designed by Parnall's designer Harold Bolas, the Panther being tested from Filton. This humpbacked two-seat naval spotter reconnaissance biplane was of single-bay layout, powered by a 230 hp Bentley BR2 rotary engine and featured a hinged folding fuselage to facilitate deck stowage. Inflatable flotation bags were located in narrow cylinders either side of the landing gear for ditching at sea, and a hydrovane was fitted to avoid nosing over if forced to descend on water.

After the Armistice, following a dispute between Avery's and the Air Ministry, an order for 150 Panthers was transferred from Parnall & Sons to Bristol's at Filton. These Panthers, built between 1919 and 1920, equipped Fleet Spotter Reconnaissance Flights aboard the aircraft carriers *Argus* and *Hermes*, and served with 205 Squadron, as well as 406, 421, 441 and 442 Fleet Spotter/Reconnaissance Flights. No doubt Parnall's Panther contributed much to the development of early carrier deck flying.

The Puffin, Bolas's first design for George Parnall, was a two-seat fighting and reconnaissance amphibian for the fleet. Of three ordered (N136–N138), the first made its initial flight on 19 November 1920 from the Isle of Grain. A two-bay biplane with folding wings, it had a 450 hp Napier Lion engine and was mounted on a large central float attached to the fuselage by sturdy struts. Land wheels, located either side of the float, retracted manually upwards on their axle via a worm and nut gear. Lateral stability was provided by two deep wing-tip floats. An unusual feature was the location of the fin and rudder below the rear fuselage to allow an unobstructed field of fire for the gunner in his rear cockpit. No production Puffins were ordered and the trio of prototypes became experimental aircraft with the Admiralty.

Parnall's Plover did enter production, albeit on a limited scale, and was a single-seat amphibious naval fighter. Of all-wooden construction it had single-bay biplane wings of equal

Designed by Harold Bolas, the Parnall Panther prototype (N91) at Filton in 1918. Note the Vickers gun on port side of nose, deleted on subsequent Panthers. (Parnall & Sons)

Odd-looking Parnall Puffin amphibian (N-138) of 1921, with 450 hp Napier Lion. Note the fin and rudder located beneath the rear fuselage, the large central float and wheels. (Parnall & Sons)

span with full span ailerons in upper and lower mainplanes. As a landplane the Plover had an orthodox cross-axle landing gear, while in amphibious configuration it featured a pair of flat-topped floats with projecting sets of wheels in pairs at the centre and rear. The whole unit was attached to the fuselage by four struts and two oleo legs.

The first Plover prototype (N160) flew early in 1923 powered by a 436 hp Bristol Jupiter IV engine, the second (N161), also with a Jupiter IV, was an amphibian and the third (N162) was a landplane with a 385 hp Armstrong Siddeley Jaguar. Following trials ten production Plovers were ordered from George Parnall (N9608–N9610 and N9702–N9708) comprising nine landplanes and only one amphibian. All were delivered in 1923, entering service with 403 and

First production Parnall Plover (N9608) at Filton, with test pilot Capt Norman MacMillan in the cockpit, c. 1923. The engine is a 436 hp Bristol Jupiter IV. (Rolls-Royce plc)

404 Fleet Fighter Flights. In spite of cleaner lines and higher maximum speed than Fairey's Flycatcher, the Plover had a short service career due to a structural weakness in the centre-section. Some Plovers went to Farnborough, Filton and Yate for test purposes.

POSSUM

In an effort to speed up postal services after the First World War, the Air Ministry proposed introducing special cargo planes with the primary role of carrying mail. Together with designs from Bristol, Boulton & Paul and Westland, George Parnall entered the Possum, of which two were built (J6862 and J6863) during 1923 and 1925 respectively, as a medium-range mailplane and military conversion of the same. Powered by one 450 hp Napier Lion, the Possum was a large triplane having two outboard propellers driven by geared shafts connected to the engine. This was mounted in a mid-fuselage position, with drives to the propellers housed in the central wing completely enclosed in a casing which ran between the leading edge and front spar. Engine cooling was via external hinged radiators, designed so that the radiating surface could be turned from the cockpit in variance with the airflow.

The Possum featured a flat-sided ply-covered fuselage of wooden construction, three unstaggered equal span wings also of wood, fabric-covered, strut and wire-braced, all incorporating ailerons. The crew of three sat in open cockpits, two forward in tandem and one aft of the wings. A conventional main landing gear was used, but instead of a tailskid a swivelling tailwheel with automatic brake was fitted.

The first Possum (J6862) made its initial flight on 19 June 1923 from Filton (George Parnall was then in the process of moving to Yate), appeared at that year's RAF Hendon Show, and on 15 April 1924 arrived at Farnborough. Building of the second Possum (J6863) was delayed, and

Parnall Possum triplane (J6862) carries exhibition No. 10 for the annual RAF Hendon Show of 1923. (Parnall & Sons)

its first flight was on 27 April 1925. It was later exhibited at the RAF Hendon Show, and test-flown at Yate a number of times by Frank Courtney. Both Possums went to the A&AEE, Martlesham, as research aircraft.

PIXIE

Responding to a £1,000 prize put up by the *Daily Mail* in 1923 for an aircraft powered by an engine of not more than 750 cc capacity, the Royal Aero Club organized a Light Aeroplane Competition at Lympne aerodrome, Kent, for October 1923. In addition the Duke of Sutherland offered £500 to the pilot who made the longest flight on one gallon of petrol.

George Parnall entered the Pixie, which was designed by Harold Bolas to have interchangeable components, with fuselage, tail unit and landing gear the same, but having alternative wings and a provision to fit small or large capacity engines with various propellers. The Pixie I monoplane had large wings and a more powerful motor for altitude tests, but with small wings and a high capacity engine for speed trials it became Pixie II. Power was a 3½ hp twin-cylinder Douglas motor-cycle engine for fuel consumption tests, and a 6 hp Douglas for altitude and speed trials. Width between the Pixie's wing roots was 2 ft, large wings having a 29 ft span, the small ones 18 ft. Pilot weight allowed was 168 lb in small wing form, and the Pixie II when loaded weighed 460 lb. The airframe was of mixed wood and metal construction, and in order to facilitate a quick powerplant change the engine mounting was separate from the fuselage. To enhance fuselage streamlining and prevent severe draughts the pilot could pull a canvas cover up to his neck with a zip-fastener!

Pixie I first flew on 13 September 1923, and in Mk II form on 4 October. At the 1923 Lympne Trials Pixie I could not equal economical consumption figures of other competitors, but flew a distance of 53.4 miles on one gallon of fuel and covered ten laps of the 12½-mile course. On 11 October the Pixie came first in the Abdulla £500 speed contest at an average of 76.1 mph; two days later it flew at 81.2 mph and on 27 October it won the speed competition in the Wakefield Prize Race at Hendon.

For the 1924 trials Bolas designed the Pixie III two-seater, with rear windscreen and capable of flying in monoplane or biplane configuration. The detachable upper wing was strut-braced, and over six feet shorter in span than the lower wing. Two Pixie IIIs were built (G-EBJG and G-EBKK) with 32 hp Bristol Cherub and 35 hp Blackburne Thrush engines respectively. At the trials both aircraft suffered mechanical problems and a spare Cherub engine was installed in

Parnall Pixie III, with a 32 hp Bristol Cherub engine and modified landing gear, on Yate airfield in 1924 prior to receiving its registration G-EBJG. (Rolls-Royce plc)

G-EBJG, which was to fly in the Grosvenor Cup Race as a Mk III monoplane. However, an earlier Pixie II monoplane (G-EBKM) appeared fitted with a 1,000 cc Blackburne engine and was favourite for the race until losing its propeller on a landing approach. The race ended with G-EBJG fifth, G-EBKM sixth and G-EBKK withdrawn with engine trouble. Two military Pixie trainers built were J7323 (later G-EBKM) and J7324, both with a 696 cc Blackburne Tomtit engine. G-EBKK went to Bristol and Wessex Aero Club, but crashed in September 1930, and G-EBKM crashed on 19 April 1939. G-EBJG went to the Midland Aircraft Preservation Society at Coventry.

PERCH AND PIKE

When Air Ministry Specification 5/24 called for a two-seat naval trainer capable of operating in alternative wheeled or floatplane configuration, with carrier deck flying training ability, George Parnall submitted the Perch. In this Bolas produced a two-bay biplane with side-by-side seating powered by a 270 hp Rolls-Royce Falcon III engine which could drive a two- or four-blade propeller. This engine was encased within a neat metal cowling with a pronounced downward

The one-off Parnall Perch (N217) in landplane form at Yate in 1927. Note the sloped cowling and four-blade propeller. (Parnall & Sons)

Parnall Pike (N202) at Yate in 1927. Note the Warren girder wing bracing and arresting 'claws' on axle to contact longitudinal carrier deck wires then in use. (Parnall & Sons)

slope on its upper surface. This afforded an excellent view forward, an asset on approach to a carrier deck, but gave the aircraft a distinct humped appearance. The Perch (N217) was an all-wooden structure with wings of equal span, and a cross-axle wheeled landing gear with oleo front legs was fitted plus tailskid. Alternatively twin-floats could be fitted to the same oleos with bracing struts added when converting to seaplane configuration.

The first flight of the Perch was on 10 December 1926 at Yate with Frank Courtney at the controls. A&AEE trials followed later, plus sea trials at the MAEE, Felixstowe. No production order was placed for the Perch and N217 remained the only one built.

George Parnall's Pike (N202) dual role naval reconnaissance aircraft, built at Yate in 1926 to Specification 1/24, could also alternate as a landplane or seaplane with a cross-axle wheeled landing gear, or twin 34 ft long floats attached by six struts. The Pike, with a 450 hp Napier Lion engine, was of mixed wood and metal construction and fabric-covered. Warren girder interplane struts were fitted eliminating flying and landing wires. The wooden tail unit was conventional, fabric-covered, and featured broad-chord horn-balanced elevators with a rounded fin and rudder. Pilots reported the Pike as having a poor view from an uncomfortable cockpit, with too great a distance between the pilot and observer/gunner. Needless to say the Pike did not reach the required Air Ministry Specification 1/24 standard.

C10/C11 AUTOGIROS

Juan de la Cierva formed the Cierva Autogiro Company in Britain during 1926, and although the main contract for production was allocated to the Hamble works of Avro, two machines, the Cierva C10 and C11 were built by George Parnall & Co. at Yate. These two machines were completed in 1927, the C10 (J9038) being an experimental military type. This aircraft had a 30 ft diameter rotor with four paddle-shaped rotor blades, the outer portions of which were built up on a ribbed aerofoil section. The fixed rotorhead, supported by struts, was located above the forward fuselage in front of the open cockpit which had no windscreen. Braced stub wings were fitted to the lower front fuselage, the conventional tail unit having horn-balanced elevators. A

Nearing completion at Yate, the Cierva C10 autogiro J9038 with 90 hp Armstrong Siddeley Genet. Note the stub wings and horn-balanced elevators. The Parnall Imp is visible in the background. (T.I. Jackson Ltd, Yate)

cross-axle landing gear and tailskid was used, the engine being a 70 hp Armstrong Siddeley Genet 1 radial with wooden propeller. Taxying trials began in April 1928, but the autogiro was badly damaged after turning on its side. Repairs were carried out and further tests made at Andover, but on 5 November J9038 crashed on take-off and was abandoned.

Parnall's C11 autogiro (G-EBOG) featured a Harold Bolas-designed airframe and was built in 1927. It had tandem open cockpits and a neat metal cowling surrounded its 120 hp ADC Airdisco engine, which drove a four-blade propeller. The C11 also crashed at Yate with Juan de la Cierva at the controls. It was repaired, updated with redesigned pylon and modified drive shaft, but eventually became an instructional airframe.

PETO

Probably the most ambitious and relatively successful attempt at conveying an aeroplane aboard a submarine for reconnaissance and spotting purposes was the diminutive Peto built by George Parnall at Yate. In response to Specification 16/24, which called for a two-seat naval seaplane capable of operating from a submarine and fitted with short span foldable wings, the Peto appeared in 1926. A compact two-seater it had staggered wings of unequal span, pronounced sweepback, full span upper ailerons, Warren girder interplane struts and twin Duralumin or Consuta floats (mahogany plywood sewn together by copper wire). Metal components and tubular struts were of stainless steel to combat corrosion by sea water, and the first Peto (N181) had a 135 hp Bristol Lucifer IV radial engine, replaced later by a 135 hp Armstrong Siddeley Mongoose.

Despite its small size the Peto proved quite manoeuvrable on water and taxied easily against wind and tide, but was unable to reach the required service ceiling of 11,000 ft. A second Lucifer-powered machine (N182) flew at Felixstowe in 1926, which was reported as handling well on calm water and in the air but tricky on a rough sea. Engine start-up was by means of a crank operated from the observer's cockpit.

A large 'M' class submarine (M2) was converted to aircraft carrier for the Peto, its 12 in gun replaced by a hangar able to withstand the pressure at M2's operating depth. This was a hemispherically rounded structure fitted with a large door and a derrick above for hoisting out or recovering the aircraft (normally launched from a catapult on the submarine's forward deck). The first men to fly the Peto from M2 were Lt C.W. Byas and Lt C. Keighley-Peach, who as naval airmen and submariners both received double rates of pay!

In 1930 the N181, after being damaged at Gibraltar, returned to Yate for repairs. Parnall's rebuilt the machine as N225 powered by a 135 hp Mongoose IIIC driving a Fairey-Reed metal propeller and fitted with improved floats, a special gear for connecting or disconnecting water rudders and main rudder, wing slots and square wing-tips. This Peto was involved in the M2 tragedy of 26 January 1932, when the submarine sank off Portland Bill with the loss of seven officers, fifty-one petty officers and ratings, and two RAF personnel. Why M2 failed to resurface is a question which has never been properly answered. Navy divers found the hangar door, the hatch from the pressure hull into the hangar and the conning tower all open. Several theories were put forward, but the true cause will probably never be known. What is certain is that the Royal Navy realized the problems involved and abandoned the concept of submarine-borne aircraft.

Parnall Peto submarine-borne biplane (N181) standing at Yate in 1926. The engine is a 135 hp Armstrong Siddeley Mongoose IIIC. It later embarked aboard the ill-fated submarine M2. (Parnall & Sons)

Peto N255 was eventually recovered from M2 (the submarine still rests on the sea-bed today after attempts to raise her were thwarted by bad weather), but so badly damaged it had to be scrapped. As for N181, after being launched from M2 off Ryde on 29 June 1930, it force-landed on Stokes Bay beach en route to Lee-on-Solent. Later withdrawn as a military machine, it was sold privately as G-AOCJ and went into storage.

IMP AND ELF

George Parnall & Co. produced two private/sporting biplanes in 1927 and 1929, these being the Imp and Elf respectively. The Imp, designed by Harold Bolas as a two-seat sporting type catering for the needs of competition pilots, was an unusual cantilever biplane with a one-piece lower wing perfectly straight in planform with no dihedral. In direct contrast the upper wing featured prominent sweepback. No bracing wires were used, interplane bracing comprising a single outward slanting wide-chord strut and centre-section cabane. All-wood affair the Imp's wing surfaces were of spruce sheeting and doped-on Egyptian cotton fabric, with a plywood-covered fuselage. An 80 hp Armstrong Siddeley Genet II was initially installed.

The sole Imp (G-EBTE) first flew in 1927 and the following year was entered in the King's Cup Air Race. Piloted by Flt Lt D.W. Bonham-Carter it came in eighth at an average speed of 109.93 mph. Later it was chosen as a test-bed for the new Pobjoy 'P' radial engine, and when it emerged at Yate updated with the Pobjoy engine the Imp had a revised front section with streamlined cowling, pointed spinner, and headrest added to the rear cockpit. During test flights from Yate the Pobjoy-powered Imp came up to expectations and the new engine looked a winner, but D.R. Pobjoy, designer of the engine, decided to build it himself and opened a factory at Hooton. Thus Parnall's hopes of producing the new engine at Yate were dashed. The Imp was later sold to Flg Off A.T. Orchard at Worthy Down in August 1933, but was scrapped that December.

The Parnall Elf of 1929 was a two-seat biplane with tandem open cockpits, the first machine (G-AAFH) having a 105 hp Cirrus Hermes I engine. Readily convertible from landplane to

The clean looking Parnall Imp two-seat sporting biplane with 80 hp Armstrong Siddeley Genet. It came eighth in the 1928 King's Cup Race averaging 109.93 mph. (Parnall & Sons)

The prototype Parnall Elf at Yate showing its wing-folding facility, with lower wing trailing edges hinged forward. It later became G-AAFH; the engine was a 105 hp Cirrus Hermes I. (Parnall & Sons)

seaplane form, the Elf featured folding wings to facilitate ground handling and Warren girder interplane bracing. Of all-wooden build the Elf was designed by Bolas to meet the growing demand at the time for a cruising speed of 100 mph which did not impair the aircraft's landing speed. Also sufficient fuel could be carried for a long distance flight up to a gross weight of 1,900 lb. There was a large luggage compartment aft of the pilot and an auxiliary locker ahead of the passenger seat.

The first of the three Elves was sold to Lord Apsley of Badminton in December 1932; this machine also flew at one time with the Cornwall Aviation Co. of St Austell. On 20 March 1934 G-AAFH suffered a fuel pump failure, crashed near Rickmansworth and was written off. Two following Elves (G-AAIO and G-AAIN) were built as Mk IIs with uprated 120 hp Cirrus Hermes II engines, improved flying controls and horn-balanced rudders. But G-AAIO crashed at Sapperton, Gloucestershire, on 13 January 1934 after fuel pump stoppage, killing both the owner, R. Hall of the Cotswold Aero Club, and his son. G-AAIN, like the original Elf, went to Lord Apsley and was stored throughout the Second World War. Partly refurbished in 1951, it was in due course acquired by Shuttleworth Trust where it was restored to flying condition, and made its first flight as such on 25 June 1980.

PIPIT

Air Ministry Specification 21/26 called for a single-seat naval fighter, and George Parnall responded with the Pipit. This was a clean looking, all-metal, fabric-covered biplane with a streamlined aluminium sheet cowling built around a 495 hp Rolls-Royce FXI liquid-cooled engine. The single-bay layout featured staggered equal span wings with ailerons fitted to the lower wings only. All four mainplanes were detachable with slackening of the flying wires obviated by jury

Prototype Parnall Pipit naval fighter (N232) under construction at Yate, c. 1927–8. (T.I. Jackson Ltd, Yate)

struts. The fuselage was a Duralumin and stainless-steel structure of oval section, and a retractable radiator was fitted which also supplied cool or heated air for the cockpit via a controllable vent.

The Pipit (N232) alternated as landplane or seaplane, being fitted in its latter form with lengthy twin floats. Armament comprised two Vickers guns firing through the propeller, with provision for an underwing bomb load. During official trials at the A&AEE in October 1928, serious tailplane flutter occurred on a steep diving test causing fracture of the tailplane spar. The Pipit went out of control and pilot Flt Lt J. Noakes tried in vain to save the machine. He was rescued from the wreck with a broken neck, but survived to become a Squadron Leader!

The second Pipit built, again as N232, had horn-balanced rudder, strut-braced tailplane and ailerons fitted to all four mainplanes. Parnall's hopes rested on this revised Pipit as its success could lead to a substantial production order. Flight tests began early in 1929, but on 17 February the second Pipit, flown by Flt Lt H.N. Pope, suffered repeat tail flutter. As the fin and rudder broke away, Pope took to his parachute, the illfated Pipit crashing on a railway embankment at Westerleigh.

Prawn

To determine the effects of engine installation in the prow of a flying boat, Parnall's one-off diminutive Prawn was built for the Air Ministry and emerged at Yate in 1930. Powered by a 65 hp Ricardo-Burt liquid-cooled engine, the Prawn (S1576) featured an all-metal hull, single open cockpit with headrest, a parasol monoplane wing, braced tailplane and outboard floats supported by N-struts and diagonals. The main problem was spray affecting the propeller, and

Parnall Prawn single-seat experimental flying boat (S1576) at Yate in 1930, with its Ricardo-Burt engine in the raised position. This gave the small diameter propeller clearance when taking off from water. (Parnall & Sons)

Bolas overcame this by mounting the engine well above the bow with a pivoting device at its rear. This enabled thrust angle to be altered by raising the whole engine to a maximum of 22°. A very small diameter four-blade propeller was fitted, its spinner forming the prow in lowered position. When raised, the engine and top-mounted radiator greatly restricted the pilot's view on take-off. After trials at Felixstowe, the Prawn remained there for a time on experimental duties.

PARASOL

In response to an Air Ministry request for a research aircraft capable of testing wing characteristics in flight, Parnall's built two Parasol high-wing monoplanes. Bolas designed these to allow the wing freedom of restricted movement in relation to the fuselage. A dynamometer, fitted to the wing struts, enabled forces acting on the wing at various angles of

Pictured at Yate airfield is 1929 is the first of two Parnall Parasol in-flight research monoplanes (K1228 and K1229) with 226 hp Armstrong Siddeley Lynx IV radial. (Parnall & Sons)

incidence to be accurately measured by dynamometer gear operated from the front cockpit by an observer. Fixing points of the wing bracing struts could be altered via a swinging frame, which provided three different incidence positions. For testing wings with slots or other high-lift devices there were alternative wing attachment points, and changes in bracing strut attachment positions could be made with comparative ease.

The Parasols (K1228 and K1229) appeared at Yate late in 1929 and were of mixed wood and metal construction. From his rear cockpit the pilot could stop the Armstrong Siddeley Lynx engine with a hydraulic lever in order to allow dynamometer readings in gliding flight. The engine was restarted by a RAE Mk II gas starter. For recording the behaviour of wool tufts stretched across the wing, a camera could be mounted atop the rear fuselage. Both Parasols were flown at Farnborough; K1229 was withdrawn in August 1936, and K1228 the following January.

G4/31

Air Ministry Specification G4/31 called for a general purpose aircraft to replace the RAF's DH9As. From George Parnall came a large all-metal, fabric-covered, two-bay biplane powered by a 690 hp Bristol Pegasus IM3, the upper wing roots incorporating a marked dihedral which produced a prominent gull wing effect. In this layout H.V. Clarke, who had replaced Bolas at Yate, used a divided wide-track landing gear for torpedo-carrying, spatted wheels and underwing bomb racks. Defensive armament comprised a forward-firing Vickers gun and a Lewis gun in the rear cockpit. Vertical tail surfaces had an odd appearance because of a very large fin, and between the wings was a cabin for personnel or stretcher cases. Parnall's G4/31, the largest and last contender in the competition, did not fly until 1935. Like the others, including Vickers Type 253 which was almost accepted, it was rejected in favour of the Vickers Wellesley monoplane bomber.

A rare photo of the Parnall G4/31 general purpose biplane at Yate prior to its metal airframe being fabric-covered. (T.I. Jackson Ltd, Yate)

The Heck Series

The prototype Hendy Heck (G-ACTC), built in 1934 by Westland Aircraft, became Parnall Heck after Hendy Aircraft Co. and Nash & Thompson bought up George Parnall to form Parnall Aircraft Ltd in 1935. This Heck first flew in 1934 as a two-seat wooden cabin monoplane with a 200 hp de Havilland Gipsy Six engine. Despite a retractable landing gear, which caused some problems and mishaps, it was decided to produce the Heck at Yate, the first of six Parnall Heck 2Cs (G-AEGH) appearing in the autumn of 1936. It contained three seats, a cabin entrance door in place of a hinged canopy, small side windows instead of the original glazing, and fixed landing gear with faired legs and spats. It was later impressed into RAF service as NF749.

The other five were G-AEGI/GJ/MR/GL and K8853, the latter being for the Air Ministry.

Built for the Air Ministry, this Parnall Heck IIC (K8853) was used at the RAE on development trials of Hurricane and Spitfire gunsights. (Parnall & Sons)

Known as the Heck III, this was the Parnall Type 382 two-seat trainer. After registration as G-AFKF it was impressed into RAF service as R9138. (Parnall & Sons)

The civil machines were impressed for war service, but G-AEGL became test-bed for the Wolseley Aries engine and was employed at the RAE. As for K8853, from 1937 it was trials aircraft for Browning machine-gun installations and reflector gunsights to be used on Hurricanes and Spitfires. Later it was the personal aircraft of Air Marshal Smart and ended its days as ground instructional airframe 3125M.

The last aircraft to bear Parnall's name, the Type 382, was to Specification T1/37, which required a two-seat trainer larger and heavier than the Miles Magister, with a 200 hp Gipsy Six engine. The new machine (JI), which emerged at Yate in 1938, was based on the Heck but had tandem open cockpits and was known as the Heck III. Although it had fixed landing gear a retractable type control facsimile was included and a blind flying hood. The first primary trainer with interconnected slots and flaps, Heck III could fly at 43 mph under power, with no danger of stalling until the angle of incidence exceeded 30°. Diving tests at 265 mph had no ill effect and little fault was found with the type, but production status was not reached. JI, registered G-AFKF, was impressed as R9138, serving with 24 Squadron for a time before ending its days as instructional airframe 3600M.

Contracts

During the 1920s George Parnall rebuilt a number of DH9As for the RAF and also produced a batch of new machines. Rebuilds at Yate were J8172–J8189, of which J8183 had its rear gun mounting removed for passenger use. New Yate-built DH9As (J8483–J8494) were completed as dual-control training aircraft.

A civil contract for Parnall's to build a Hendy 302 two-seat sporting monoplane for Hendy Aircraft Co. resulted in a wooden design by Basil Henderson. With enclosed tandem cockpits, fixed divided landing gear and a 105 hp Cirrus Hermes I, the Hendy 302 (G-AAVT) appeared at Yate in 1929. It was piloted by Capt E.W. Percival in the 1930 King's Cup Race achieving 121.51 mph, and in the 1931 Heston to Newcastle Race covered the course at an average speed of 145 mph. It went to Cirrus Hermes Engineering for use as a test-bed, and later passed to Carill S. Napier. After rebuild as the Hendy 302A, G-AAVT flew in several races and was a test-bed for the 150 hp Cirrus Hermes II, before being finally withdrawn in 1938.

The Parnall-built Hendy 302 (G-AAVT) following its 1933 update as the 302A, with revised cockpit canopy, wheel spats and a 130 hp Cirrus Hermes IV. (The Riding Collection)

F.G. 'Freddie' Miles in the cockpit of his Parnall-built Miles Satyr with Pobjoy 'R' engine. (Author's collection)

Parnall-built Percival Gull Four (G-ABUV), with 130 hp Cirrus Hermes IV, which crashed on 2 November 1936. (Author's collection)

Another one-off contract awarded to George Parnall was to build a small sporting wooden biplane for F.G. Miles. Named Satyr, this aircraft was purely an aerobatic and exhibition mount for Miles' personal use and featured a single open cockpit with a small windscreen, faired in headrest, a 75 hp Pobjoy 'R' radial, well-staggered wings divided by single I broad-chord interplane struts, conventional tail unit and cross-axle V-type landing gear with large diameter wheels. Registered G-ABVG the Satyr first flew at Yate in August 1932 piloted by F.G. Miles, who later flew it from Yate to Shoreham. In May 1933 the Satyr was sold to the Hon Mrs

Victor Bruce and was flown until September 1936, when it crashed and was written off.

The most significant civil contract awarded to George Parnall was to produce twenty-three Percival Gull Four three-seat cabin monoplanes. The prototype (G-ABUR) had been built by Capt E. Percival at Maidstone powered by a 160 hp Napier Javelin, and was faster than some contemporary fighters. Being all-wood the Gull Four was an ideal subject for George Parnall with its expertise in wooden airframes. Alternative engine to the Cirrus Hermes IV normally fitted to Gull Fours was either a 130 hp Gipsy Major, or 160 hp Napier Javelin III. From those production Gull Fours built at Yate between 1933 and 1934, a number found their way overseas to Australia, Brazil, France, India and the Netherlands. One Parnall-built Gull Four (G-ACJV) was used by Sir Charles Kingsford Smith on his record England to Australia flight in October 1933. Named *Miss Southern Cross* it became VH-CKS.

APPENDIX I
Technical Data

Principal Gloster, Bristol, Parnall Aircraft

1. *Gloster Aircraft Co. Ltd*

MARS 1 'BAMEL' (original version)
Type: Single-seat biplane racer.
Powerplant: One 450 hp Napier Lion.
Performance: Max speed, 196.6 mph. Climb, 4¼ min to 10,000 ft.
Weights: Empty, 1,890 lb. Loaded, 2,500 lb.
Dimensions: Span, 22 ft. Length, 23 ft. Height, 9 ft 4 in. Wing area, 205 sq ft.

GLOSTER IIIA
Type: Single-seat biplane racing seaplane.
Powerplant: One 700 hp Napier Lion VII.
Performance: Max speed, 225 mph (SL).
Weights: Empty, 2,028 lb. Loaded, 2,687 lb.
Dimensions: Span, 20 ft. Length, 26 ft 10 in. Height, 9 ft 8 in. Wing area, 152 sq ft.

GLOSTER GREBE II
Type: Single-seat biplane fighter.
Powerplant: One 400 hp Armstrong Siddeley Jaguar IV radial.
Performance: Max speed, 152 mph (SL). Climb, 23 min to 20,000 ft. Service ceiling, 23,000 ft.
Weights: Empty, 1,720 lb. Loaded, 2,614 lb.
Dimensions: Span, 29 ft 4 in. Length, 20 ft 3 in. Height, 9 ft 3 in. Wing area, 254 sq ft.
Armament: Two .303 in Vickers machine-guns.

GLOSTER GAMECOCK
Type: Single-seat biplane fighter.
Powerplant: One 425 hp Bristol Jupiter VI radial.
Performance: Max speed, 155 mph at 5,000 ft. Climb, 7.6 min to 10,000 ft; 20 min to 20,000 ft. Service ceiling, 22,000 ft.
Weights: Empty, 1,930 lb. Loaded, 2,863 lb.
Dimensions: Span, 29 ft 9½ in. Length, 19 ft 8 in. Height, 9 ft 8 in. Wing area, 264 sq ft.
Armament: Two .303 in Vickers machine-guns.

GLOSTER GAUNTLET I
Type: Single-seat biplane fighter.
Powerplant: One 645 hp Bristol Mercury VIS2 radial.
Performance: Max speed, 230 mph at 15,800 ft. Climb, 2,300 ft/min. Service ceiling, 33,500 ft. Range, 460 miles.
Weights: Empty, 2,775 lb. Loaded, 3,970 lb.

Dimensions: Span, 32 ft 9½ in. Length, 26 ft 2 in. Height, 10 ft 4 in. Wing area, 315 sq ft.
Armament: Two .303 in Vickers machine-guns.

GLOSTER GLADIATOR I
Type: Single-seat biplane fighter.
Powerplant: One 840 Bristol Mercury IX radial.
Performance: Max speed, 253 mph at 14,500 ft. Climb, 2,300 ft/min. Service ceiling, 33,000 ft.
Weights: Empty, 3,450 lb. Loaded, 4,750 lb.
Dimensions: Span, 32 ft 3 in. Length, 27 ft 5 in. Height, 10 ft 4 in. Wing area, 323 sq ft.
Armament: Four .303 in Browning machine-guns.

GLOSTER E28/39
Type: Single-seat jet-propelled research monoplane.
Powerplant: One 860 lb thrust (st) Power Jets W1, or 1,160 lb st W1A, or 1,700 lb st W2/500 turbojet.
Performance: 466 mph at 10,000 ft. Climb, 22 min to 30,000 ft. Service ceiling, 32,000 ft.
Weights: Empty, 2,886 lb. Loaded, 3,748 lb.
Dimensions: Span, 29 ft. Length, 25 ft 3¾ in. Height, 9ft 3in. Wing area, 146.5 sq ft.

GLOSTER METEOR F8
Type: Single-seat jet-propelled interceptor fighter.
Powerplant: Two 3,600 lb st Rolls-Royce Derwent 8 turbojets.
Performance: Max speed, 598 mph at 10,000 ft. Climb, 6,950 ft/min. Range, 980 miles. Service ceiling, 45,000 ft.
Weights: Empty, 10,626 lb. Loaded (max with ventral and drop tanks), 19,100 lb.
Dimensions: Span, 37 ft 2 in. Length, 44 ft 7 in. Height, 13 ft 10 in. Wing area, 350 sq ft.
Armament: Four 20 mm Hispano cannon.

GLOSTER JAVELIN FAW9
Type: Two-seat jet-propelled all-weather fighter.
Powerplant: Two 11,000 lb st (13,390 lb with reheat) Bristol Siddeley Sapphire 203/204 turbojets.
Performance: Max speed, 620 mph at 40,000 ft. Climb, 5.6 min to 40,000 ft. Range, 930 miles. Service ceiling, 52,000 ft.
Weights: Loaded (normal), 38,100 lb; with full ventral tanks, 43,165 lb.
Dimensions: Span, 52 ft. Length, 56 ft 9 in. Height, 16 ft 3 in. Wing area, 928 sq ft.

2. *Bristol Aeroplane Co. Ltd*

BRISTOL SCOUT D
Type: Single-seat scouting biplane.
Powerplant: One 80 hp Le Rhone 9C rotary.
Performance: Max speed, 92.7 mph (SL). Climb, 7 min to 5,000 ft. Service ceiling, 15,500 ft.
Weights: Empty, 760 lb. Loaded, 1,250 lb.
Dimensions: Span, 24 ft 7 in. Length, 20 ft 8 in. Height, 8 ft 6 in. Wing area, 198 sq ft.
Armament: One .303 in Lewis machine-gun above centre-section (some later with .303 in Vickers via interrupter gear).

APPENDIX I

BRISTOL F2B FIGHTER MK III (post-First World War)
Type: Two-seat Army Cooperation biplane/dual-control trainer.
Powerplant: One 280 hp Rolls-Royce Falcon III.
Performance: Max speed, 110 mph (SL). Climb, 838 ft/min. Service ceiling 20,000 ft.
Weights: Empty, 2,150 lb. Loaded, 3,250 lb.
Dimensions: Span, 39 ft 4 in. Length, 25 ft 10 in. Height, 9 ft 9 in. Wing area, 406 sq ft.
Armament: One forward-firing .303 in Vickers gun; one (or two yoked) .303 in Lewis gun(s) in rear cockpit. Two 112 lb or twelve 20 lb Cooper bombs.

BRISTOL M1C
Type: Single-seat monoplane fighter.
Powerplant: One 110 hp Le Rhone 9J rotary.
Performance: Max speed, 111.5 mph at 10,000 ft. Climb, 10 min 10 sec to 10,000 ft. Service ceiling, 20,000 ft.
Weights: Empty, 896 lb. Loaded, 1,348 lb.
Dimensions: Span, 30 ft 9 in. Length, 20 ft 5½ in. Height, 7 ft 9½ in. Wing area, 145 sq ft.
Armament: One .303 in Vickers machine-gun.

BRISTOL BULLDOG IIA
Type: Single-seat day/night fighter.
Powerplant: One 440 hp Bristol Jupiter VIIF/VIIF.P.
Performance: Max speed, 174 mph at 10,000 ft. Climb, 14½ min to 20,000 ft. Service ceiling, 29,300 ft.
Weights: Empty, 2,412 lb. Loaded, 3,530 lb.
Dimensions: Span, 33 ft 10 in. Length, 25 ft 2 in. Height, 8 ft 9 in. Wing area, 306.5 sq ft.
Armament: Two .303 in Vickers machine-guns. Provision for four 20 lb bombs on rack beneath port lower wing.

BRISTOL BOMBAY
Type: Heavy bomber/transport and troop carrier monoplane.
Powerplant: Two 1,010 hp Bristol Pegasus XXII radials.
Performance: Max speed, 192 mph at 6,500 ft. Climb, 750 ft/min. Range (normal), 880 miles; maximum (with extra tankage), 2,230 miles. Service ceiling, 25,000 ft.
Weights: Empty, 13,800 lb. Loaded, 20,000 lb.
Dimensions: Span, 95 ft 9 in. Length, 69 ft 3 in. Height, 19 ft 11 in. Wing area, 1,340 sq ft.
Armament: Four Vickers 'K' guns; one each in nose, tail, port and starboard beam positions. Maximum 2,000 lb bomb load.

BRISTOL BLENHEIM IV
Type: Three-seat light bomber.
Powerplant: Two 905 hp Bristol Mercury XV radials.
Performance: Max speed, 266 mph at 11,800 ft. Climb, 1,500 ft/min. Range, 1,460 miles. Service ceiling, 22,000 ft.
Weights: Empty, 9,790 lb. Loaded, 14,400 lb.
Dimensions: Span, 56 ft 4 in. Length, 42 ft 7 in. Height, 9 ft 10 in. Wing area, 469 sq ft.
Armament: Five .303 in Browning machine-guns; one in port wing, two in undernose blister firing aft, two in dorsal turret. Bomb load, 1,000 lb internal and 320 lb external.

BRISTOL BEAUFORT I
Type: Four-seat torpedo-bomber.
Powerplant: Two 1,130 hp Bristol Taurus VI radials.
Performance: Max speed, 265 mph at 6,000 ft. Range (normal), 1,035 miles. Service ceiling, 16,500 ft.
Weights: Empty, 13,107 lb. Loaded, 21,228 lb.
Dimensions: Span, 57 ft 10 in. Length, 44 ft 7 in. Height, 12 ft 5 in. Wing area, 503 sq ft.
Armament: Four .303 in Browning machine-guns, two each in nose and dorsal turrets (some with additional gun later firing aft from undernose blister plus two beam guns). One 1,605 lb 18 in torpedo or up to 1,500 lb of bombs.

BRISTOL BEAUFIGHTER VIF
Type: Two-seat night fighter.
Powerplant: Two 1,670 hp Bristol Hercules VI or XVI radials.
Performance: Max speed, 333 mph at 15,600 ft. Climb, 7.8 min to 15,000 ft. Range, 1,470 miles. Service ceiling, 26,500 ft.
Weights: Empty, 14,600 lb. Loaded, 21,600 lb.
Dimensions: Span, 57 ft 10 in. Length, 41 ft 8 in. Height, 15 ft 10 in. Wing area, 503 sq ft.
Armament: Four nose-mounted 20 mm cannon and six wing-mounted .303 in Browning machine-guns.

BRISTOL BRIGAND B1
Type: Three-seat light ground-attack bomber.
Powerplant: Two 2,470 hp Bristol Centaurus 57 radials.
Performance: Max speed, 358 mph at 16,000 ft. Climb, 1,500 ft/min. Range (normal), 1,980 miles. Service ceiling, 26,000 ft.
Weights: Empty, 25,598 lb. Loaded, 39,000 lb.
Dimensions: Span, 72 ft 4 in. Length, 46 ft 5 in. Height, 17 ft 6 in. Wing area, 718 sq ft.
Armament: Four nose-mounted 20 mm Hispano cannon. Up to 2,000 lb of ordnance (bombs or rocket projectiles).

BRISTOL TYPE 170 FREIGHTER MK I
Type: High-wing cargo or passenger (32-seater) monoplane.
Powerplant: Two 1,675 hp Bristol Hercules 632 radials.
Performance: Max speed, 236 mph at 8,000 ft. Climb, 1,135 ft/min. Range, 300 miles. Ceiling, 22,000 ft.
Weights: Empty, 23,482 lb. Loaded, 36,500 lb.
Dimensions: Span, 98 ft. Length, 68 ft 4 in. Height, 21 ft 8 in. Wing area, 1,405 sq ft.

BRISTOL BRITANNIA SERIES 310
Type: Long-range commercial transport.
Powerplant: Four 4,120 ehp Bristol Siddeley Proteus 755 turbo-props.
Performance: Max speed, 397 mph. Service ceiling, 24,000 ft. Range (max payload), 4,268 miles.
Weights: Empty, 82,537 lb. Loaded (max take-off), 185,000 lb.
Dimensions: Span, 142 ft 3 in. Length, 124 ft 3 in. Height, 37 ft 6 in. Wing area, 2,075 sq ft.

APPENDIX I

BAC (BRISTOL)/AEROSPATIALE CONCORDE (production)
Type: Long-range supersonic commercial transport aircraft.
Powerplant: Four 38,050 lb st (with reheat) Rolls-Royce/SNECMA Olympus 593 Mk 602 turbojets.
Performance: Max cruising speed, 1,450 mph (Mach 2.2) at 54,500 ft. Climb, 5,000 ft/min. Maximum range at 1,350 mph cruise (Mach 2.05) with reserve fuel and 17,000 lb payload, 4,400 miles. Maximum payload range, 4,020 miles at 1,350 mph (Mach 2.05) at 54,500 ft
Weights: Empty, 169,000 lb. Loaded (max take-off), 385,810 lb.
Dimensions: Span, 84 ft. Length, 203 ft 8¾ in. Height, 39 ft 10¼ in. Wing area, 3,856 sq ft.

3. George Parnall & Co.

PARNALL PLOVER
Type: Single-seat shipboard biplane fighter.
Powerplant: One 436 hp Bristol Jupiter IV, or 385 hp Armstrong Siddeley Jaguar radial.
Performance: Max speed, 142 mph (SL). Climb, 25 min 13 sec to 20,000 ft. Service ceiling, 23,000 ft.
Weights (Jupiter engine): Empty, 2,035 lb. Loaded, 2,984 lb.
Dimensions: Span, 29 ft. Length, 23 ft. Height, 12 ft. Wing area, 306 sq ft.
Armament: Two .303 in Vickers machine-guns.

PARNALL PIXIE I AND II
Type: Single-seat sporting or military training monoplane.
Powerplant: One 3½ hp (500 cc) or 6 hp (736 cc) Douglas air-cooled flat-twin.
Performance: Max speed (736 cc Douglas) 81.2 mph timed against 35 mph headwind. Constructor's estimated max speed with 1,000 cc Blackburne (29 ft wing), 90 mph. Ceiling, 18,000 ft. Estimated max speed with 1,000 cc Blackburne (18 ft wing), 105 mph. Ceiling, 14,500 ft.

PARNALL PETO
Type: Two-seat submarine-borne naval reconnaissance twin-float biplane seaplane.
Powerplant: One 135 hp Bristol Lucifer IV 3-cylinder radial, or 135 hp Armstrong Siddeley Mongoose IIIC 5-cylinder radial.
Performance: Max speed (Mongoose IIIC), 113 mph (SL). Climb, 600 ft/min. Service ceiling (Lucifer IV), 9,550 ft.
Weights: Empty, 1,300 lb. Loaded, 1,950 lb.
Dimensions: Span, 28 ft 5 in. Folded, 8 ft. Length, 22 ft 6¼ in. Height, 8 ft 11 in. Wing area, 174 sq ft.

PARNALL ELF II
Type: Two-seat private-owner and sporting biplane.
Powerplant: One 120 hp Cirrus Hermes II air-cooled inline.
Performance: Max speed, 116 mph (SL). Climb, 800 ft/min. Ceiling, 16,000 ft. Range, 400 miles.
Weights: Empty, 900 lb. Loaded, 1,700 lb.
Dimensions: Span, 31 ft 3½ in. Length, 22 ft 10½ in. Height, 8 ft 6 in. Wing area, 195 sq ft.

PARNALL HECK 2C

Type: Three-seat private cabin monoplane and (impressed) military communications monoplane.
Powerplant: One 200 hp de Havilland Gipsy Six inverted inline air-cooled.
Performance: Max speed, 185 mph. Climb, 1,000 ft/min. Service ceiling, 16,700 ft. Range, 605 miles.
Weights: Empty, 1,750 lb. Loaded, 2,700 lb.
Dimensions: Span, 31 ft 6 in. Length, 26 ft 1½ in. Height, 8 ft 6 in. Wing area, 105.2 sq ft.

PARNALL TYPE 382 HECK III

Type: Two-seat elementary military training monoplane.
Powerplant: One 200 hp de Havilland Gipsy Six.
Performance: Max speed, 155 mph (SL). Climb, 780 ft/min. Range, 620 miles. Service ceiling, 17,000 ft.
Weights: Empty, 1,655 lb. Loaded, 2,450 lb.
Dimensions: Span, 33 ft 6 in. Length, 28 ft 8 in. Height, 7 ft 9 in. Wing area, 155 sq ft.

APPENDIX 2
Glossary

A&AEE	Aircraft & Armament Experimental Establishment
AAM	Air-to-Air Missile
AASF	Advanced Air Striking Force
AEF	Air Experience Flight
AFC	Australian Flying Corps
AFU	Advanced Flying Unit
AHB	Air Historical Branch
AITD	Advanced Instrumentation Techniques Department
AONS	Air Observers Navigation School
APU	Aircraft Preparation Unit
ASU	Aircraft Storage Unit
ATA	Air Transport Auxiliary
AuxAF	Auxiliary Air Force
BAC	British Aircraft Corporation
BAe	British Aerospace
BEA	British European Airways
BEF	British Expeditionary Force
BOAC	British Overseas Airways Corporation
Capt	Captain
CFS	Central Flying School
C of A	Certificate of Airworthiness
D/F	Direction Finding
E&RFTS	Elementary & Reserve Flying Training School
FAA	Fleet Air Arm
FIS	Flying Instructors School
FLS	Fighter Leader School
Flt Lt	Flight Lieutenant
Flg Off	Flying Officer
FTD	Flight Test Department
FTS	Flying Training School
GAC	Gloucestershire Aircraft Co./Gloster Aircraft Co.
Gp Capt	Group Captain
hp	horse power
LTW	Lyneham Transport Wing
MAEE	Marine Aircraft Experimental Establishment
MAP	Military Aircraft Photographs
ME	Middle East
MoD	Ministry of Defence
MoS	Ministry of Supply
mph	miles per hour
MU	Maintenance Unit

GLOSSARY

N/A	Not Available
NZPAF	New Zealand Permanent Air Force
OAPU	Overseas Aircraft Preparation Unit
OTU	Operational Training Unit
(P)	Pilot
Plt Off	Pilot Officer
RAAF	Royal Australian Air Force
RAE	Royal Aircraft Establishment
RAF	Royal Air Force
R.A.F.	Royal Aircraft Factory
RAFC	Royal Air Force College
RAN	Royal Australian Navy
RCAF	Royal Canadian Air Force
RCU	Radar Calibration Unit
RFC	Royal Flying Corps
RFS	Reserve Flying School
RNAS	Royal Naval Air Service
RNZAF	Royal New Zealand Air Force
rpm	revolutions (revs) per minute
SAM	Surface-to-Air Missile
SAL	Scottish Aviation Ltd
SBAC	Society of British Aircraft Constructors
SFTS	Service Flying Training School
SL	Sea Level
SOC	Struck Off Charge
Sqn Ldr	Squadron Leader
SSA	Single-seat Armoured
SST	Supersonic Transport
TEU	Tactical Exercise Unit
UAS	University Air Squadron
UK	United Kingdom
USAF	United States Air Force
VHF	Very High Frequency
Wg Cdr	Wing Commander
YDOHP	Yate District Oral History Project

Bibliography

A History of Passenger Aircraft; W. Sweetman. Hamlyn, 1980.
Aircraft of the 1914–1918 War; O.G. Thetford/E.J. Riding. Harborough, 1954.
Aircraft of the RAF Since 1918; O. Thetford. Putnam (8th ed.), 1988.
Aircraft of the Fighting Powers Vols 1–VII; O.G. Thetford/E.J. Riding. Harborough, 1941–6.
An Elementary Course in the Construction of Aircraft, Sec. 4, Ch. VI, Subsidiary structures: The undercarriage; M. Langley. New Era, 1943.
Armstrong Whitworth Aircraft Since 1913; O. Tapper. Putnam, 1973.
Armament of British Aircraft 1909–1939; H.F. King. Putnam, 1971.
Bristol Blenheim; C. Bowyer. Ian Allan, 1984.
Bristol Brabazon: Press Office; Bristol Aeroplane Co., 1949.
Bristol: An Aircraft Album; J.D. Oughton. Ian Allan, 1973.
British Civil Aircraft Since 1919 (3 vols); A.J. Jackson. Putnam, 1973–4.
British Research and Development Aircraft; R. Sturtivant. Haynes, 1990.
British Naval Aircraft Since 1912; O. Thetford. Putnam (5th ed.), 1982.
British Aeroplanes 1914–18; J.M. Bruce. Putnam, 1957.
British Isles Civil Register; R.J. Hoddinott & D.S. Seex. Laas International, 1980.
Combat Aircraft of the World; J.W.R. Taylor. Ebury Press & Michael Joseph, 1969.
Concise Guide to British Aircraft of World War II; D. Mondey. Temple Press, 1984.
Concorde: Flight into the Future; Terry Hughes and John Costello. Macmillan, 1972.
De Havilland Aircraft Since 1909; A.J. Jackson. Putnam, 1962.
Duel in the skies over Windrush; J. Rennison. Cotswold Life c. 1988.
Famous Fighters of the Second World War; W. Green. Macdonald (1st ed.), 1957.
Famous Fighters of the Second World War; W. Green. Macdonald (2nd Ser.), 4th Imp., 1969.
Famous Bombers of the Second World War; W. Green. Macdonald (2nd Ser.) 2nd Imp., 1962.
Gloster Aircraft Since 1917; D.N. James. Putnam, 1971.
Gloster Javelin; M. Allward. Ian Allan, 1983.
Hawker Aircraft Since 1920; F.K. Mason. Putnam, 1961.
History of RAF Little Rissington 1938–1960; PRO, RAF/CFS, Little Rissington, c. 1961.
Interceptor; J. Goulding. Ian Allan, 1986.
Jane's All The World's Aircraft 1919; C.G. Grey. David and Charles (reprint), 1969.
Jane's All The World's Aircraft 1938; C.G. Grey and Leonard Bridgman. David and Charles (reprint), 1972.
Meteor; B. Philpott. Patrick Stephens, 1986.
No. 5 Maintenance Unit, RAF Kemble, 1939–1979; PRO, RAF, Kemble, c. 1980.
Parnall Aircraft Since 1914; K.E. Wixey. Putnam, 1990.
Parnall's In Memoriam; Yate District Oral History Project, 1990.
RAF Squadrons; Wg Cdr C.G. Jefford. Airlife, 1988.
RAF Aircraft J1–J9999; D. Thompson and R. Sturtivant. Air Britain, 1987.
RAF Aircraft K1000–K9999; J.J. Halley. Air Britain, 1976.
RAF Aircraft L1000–L9999; J.J. Halley, Air Britain, 1979.
RAF News, 20 Feb–5 March 1987.
Rotol: The History of an Airscrew Company 1937–1960; B. Stait. Alan Sutton, 1990.
Supersonic Age; British Aircraft Corporation and Aerospatiale. Winter 1975/76.

The Speed Seekers; T.G. Foxworth. Macdonald & Jane's, 1975.

The Schneider Trophy Contests (1913–1931); Ellison Hawks. Real Photographs Co. publication, 1945.

The Gloster Gladiator; F.K. Mason. Macdonald, 1964.

The Gloster Meteor; E. Shacklady. Macdonald, 1962.

The Hawker Hurricane; F.K. Mason. Macdonald, 1962.

The Book of Bristol Aircraft; D.A. Russell. Harborough, 1946.

The Aeroplane, 19 October 1956.

The Bristol Aeroplane Quarterly, Vol. 2, No. 5, Summer 1957.

The *Gloucester Citizen, Gloucestershire Echo, Gloucester Journal, Gloucester News*: various articles/letters, *c.* 1970s/1980s.

Treble One: The Story of No. 111 Squadron, RAF; Flg Off R.P.D. Sands, 1957.

Vickers Aircraft Since 1908; C.F. Andrews and E.B. Morgan. Putnam (revised ed.), 1988.

Whittle: The True Story; J. Golley. Airlife, 1987.

Index

AIRCRAFT AND ENGINES

ABC Dragonfly engine, 18, 22
ADC Airdisco engine, 125
Airship R33, 28
Airspeed Horsa, 12, 13, 14, 61
    Oxford, 10, 14, 15, 62
Alvis leonides engine, 106, 108
Anzani engine, 77
Armstrong Siddeley engines, 26, 27, 28, 29, 39, 40,
    43, 46, 66, 68, 69, 94, 120, 125, 126, 127,
    130, 131, 136
Armstrong Whitworth Albemarle, 10, 12, 60, 61
    Argosy, 4
    Atlas, 87
    Siskin, 27, 29, 36, 37, 92, 114
    Whitley, 4, 10, 12
Avro 504, 15, 26, 72, 115, 707, 66
    Anson, 9, 10, 11, 62, 73
    Lancaster, 12, 119
    Lincoln, 76, 101, 119
    Shackleton, 13
    Vulcan, 66

BAC/Aerospatiale Concorde, 11, 12, 111, 112,
    113, 140
BAe Hawk, 13
Beardmore engine, 82
Bell XP-59 Airacomet, 62
Bentley BR2 rotary engine, 18, 26, 119
Blackburn Shark, 45, 46
    Botha, 119
Blackburne engines, 25, 122, 123
Boeing B-47, 11
    B-52, 11
    KC-135, 11
    707, 108
Boulton Paul Defiant, 10, 15, 57
    P111, 66
    Sidestrand, 95
Breda 15 monoplane, 37
Bristol engines:
    Aquila, 57, 92
    Centaurus, 100, 104, 106, 139
    Cherub, 87, 122
    Hercules, 9, 61, 99, 104, 139
    Jupiter, 29, 30, 32, 33, 34, 35, 36, 37, 38, 39,
        40, 42, 43, 83, 84, 87, 88, 89, 92, 93, 94,
        120, 136, 138, 140
    Lucifer, 85, 126, 140

Mercury, 30, 36, 38, 39, 42, 43, 47, 48, 50, 89,
    90, 91, 92, 96, 136, 137, 138
Neptune, 89
Orion, 35
Pegasus, 36, 46, 74, 89, 95, 131, 138
Perseus, 36, 90
Proteus, 104, 105, 106, 107, 139
Taurus, 57, 90, 98, 139
Titan, 89
Bristol Siddeley Sapphire, 137
    Thor ramjet, 113
Bristol Babe, 83
    Baby, 79
    Badger, 82, 83, 94
    Badminton, 88
    Bagshot, 95
    Beaufort, 12, 73, 75, 98, 99, 139
    Beaufighter, 57, 73, 75, 99, 100, 101, 139
    Beaver, 88,
    Belvedere, 76, 108, 109
    Berkeley, 86, 87
    Bisley, 97, 98
    Blenheim, 10, 12, 49, 50, 57, 72, 73, 76, 95, 96,
        97, 100, 138
    Bloodhound (Type 84), 87
    Bloodhound SAM, 76, 113, 114
    Boarhound, 87
    Bolingbroke, 96, 98
    Bombay, 95, 96
    Boxkite, 71, 76, 77
    Brabazon, 73, 76, 101, 102, 103
    Braemar, 82
    Brandon, 84
    Brigand, 100, 101, 139
    Britannia, 76, 106, 107, 108, 139
    Brownie, 87, 88
    Buckingham, 100, 101
    Buckmaster, 100
    Bulldog, 29, 30, 32, 42, 43, 73, 92, 93, 95, 138
    Bullet, 84
    Bullfinch, 84
    Bullpup, 93
    Burney Hydroplane, 78, 79
    Challenger-England, 76
    Coanda aeroplanes, 77, 78
    Fighter (F2A/B), 16, 34, 81, 82, 89, 138
    Freighter, 76, 104, 105, 139
    GE1/2/3 Biplanes, 77

INDEX

Glider, 76
High Altitude Monoplane, 89, 90
Laboratory Biplane, 87
MIA/B/C Monoplanes, 82, 83, 138
MR1 Biplane, 82
Prier Monoplanes, 76, 77
Pullman, 83,
SSA/S2A Biplanes, 82
Scouts, 15, 79, 80, 137
Seely, 84
Single-seat Monoplane, 76
Sycamore, 76, 105, 106
Ten-Seater, 84, 85
Tourer, 82, 84
Tractor Biplane, 76
Tramp, 83
Twin Tractor Biplane, 82
Type 'T', 76
Type 72 Racer, 85
Type 73 Taxiplane, 85, 86
Type 83 Trainer, 86, 87
Type 101, 88
Type 109 Long-Range, 86, 87
Type 110A, 73, 89
Type 118, 89
Type 120, 89
Type 123, 91
Type 133, 91
Type 143, 92
Type 146, 90
Type 148, 90, 91
Type 155, 61,
Type 188, 109, 110
Type 221, 110, 111

Canadair CL-13 Sabre, 13
Canton-Unne engine, 79
Caproni Ca 133, 44
    Ca 310, 49
Carden engine, 25
Caudron biplane, 3
Cirrus Hermes engine, 127, 128, 133, 135, 140
Clerget engine, 77, 80, 82
Cody 'Flying Bus', 2
Cosmos engines, 73, 80, 82, 94
Curtiss R3C2 Racer, 22

De Havilland engines, 62, 110, 132, 133, 135, 141
    De Havilland DH6, 15, 16, 17
    DH9/9A, 33, 34, 39, 72, 89, 116, 133
    DH66, 39, 44
    DH67, 39
    DH72, 37
    DH77, 37
    Dove, 8

Dragon Rapide, 9
Chipmunk, 11, 13, 73
Mosquito, 15, 57, 100
Tiger Moth, 9, 72, 73
Vampire, 11, 13
Daimler engine, 77
Douglas Dakota, 13
    Thor Ballistic Missile, 113
Douglas (UK) engines, 122, 140

English Electric Canberra, 13
    Lightning, 70

Fairchild F-27, 4
Fairey Battle, 10
    Flycatcher, 121
    Gannet, 5
    Hamble (Sopwith) Baby, 115
    Rotodyne, 4
    Swordfish, 45
Farman biplanes, 15,16
Fiat CR32/42, 49
Focke-Wulf Condor, 57
Fokker Friendship, 4
    FV11b-3M, 38
Folland 43/37, 9
    Gnat, 11, 13

General Aircraft Hotspur, 12, 15
    Monospar, 37
Gloster I Racer, 20, 21
    Gloster II, 21, 31
    Gloster III, 21, 22, 23, 136
    Gloster IV, 23, 24, 35
    Gloster VI, 23, 24
    AS31, 39, 40, 41
    Bamel, 18, 19, 20, 136
    E1/44, 58, 60
    E5/42, 60
    E28/39, 58, 59, 62, 137
    FS/36, 45
    F5/34, 47, 48
    F9/37, 57, 58
    F18/37, 59
    F153D, 70
    Gambet, 32, 33
    Gamecock, 29, 30, 31, 32, 33, 92, 95, 136
    Gannet, 25, 26
    Gauntlet, 6, 42, 43, 44, 45, 48, 52, 93, 136
    Gladiator, 6, 7, 10, 44, 47, 48, 49, 50, 51, 52, 72, 93, 137
    Gnatsnapper, 38, 39
    Goldfinch, 32
    Goral, 34
    Gorcock, 31

INDEX

Goring, 34, 35, 36
Grebe, 18, 26, 27, 29, 30, 31, 136
Grouse, 26, 27
Guan, 31, 32
Javelin, 41, 62, 66, 68, 69, 70, 137
Mars, 18, 26, 95, 136
Meteor, 4, 5, 10, 11, 63, 64, 65, 66, 68, 119, 137
Monospar SS1, 37
Nighthawk (Thick-Wing), 26, 27
Nightjar, 18
Sparrowhawk, 18, 26, 32
TC33, 41, 42
TSR38, 45
Gnome engines, 3, 30, 73, 76, 77, 78, 79, 80, 94
Gosling Rocket-Boost Motor, 113
Grumman Gulfstream, 4

Halford Turbojet, 60, 62
Handley Page 42/45, 94
    Halifax, 9, 11, 61
    Hampden, 15
    Hampstead, 95
    Hare, 35
    Hinaidi/Clive, 95
Hawker Audax, 10, 14, 52, 114
    Demon, 117, 119
    Fury, 10, 43, 93
    Hardy, 52, 54
    Harrier, 35
    Hart, 10, 13, 52
    Hawfinch, 92
    Henley, 53, 54
    Hind, 12
    Horsley, 34
    Hunter, 13
    Hurricane, 9, 10, 12, 14, 44, 47, 48, 49, 50, 52, 53, 54, 55, 57, 58, 72, 132, 133
    Sea Fury, 5
    Tempest II, 114
    Typhoon, 9, 10, 12, 15, 56, 57, 59
    Woodcock, 95
Hawker Siddeley 748/Andover 8, 10, 13
Heinkel He 111, 10, 49, 50 119
Hendy 302, 133
Heston Phoenix, 6, 7
Hispano-Suiza engine, 81, 82
Hunting Jet Provost, 11, 13
    Sea Prince, 13

Junkers Ju 88, 52

Le Rhone engine, 78, 80, 82, 137, 138
Liberty engine, 33, 82
Lockheed C–130 Hercules, 5, 11

Lorraine-Dietrich engine, 40

Macchi M33 Racer, 22
    M52 Racer, 23
Messerschmitt Bf 110, 50, 72
Metropolitan Vickers engine, 62
Miles Martinet, 10, 15
    Master, 10, 14, 15
    Satyr, 134, 135

Nakajima Kotobuki engine, 33
    Licence-built Jupiter, 92
Napier engines, 4, 19, 20, 21, 22, 23, 31, 32, 34, 37, 57, 84, 106, 108, 119, 121, 124, 135, 136
Navy Type 3 carrier fighter (Japan), 33
Nieuport LC1 Nieuhawk, 18
    Nighthawk, 16, 18, 26
North American Harvard, 10
    Mustang, 15
    Sabre, 110

Palmer Aeroplanes, 3, 4
Parnall (Bolas) 'Bodyless Monoplane', 116
    C10/C11 Autogiros, 124, 125
    Elf, 116, 117, 127, 128, 140
    G4/31, 119, 131
    Heck 2C and III, 132, 133, 141
    Imp, 116, 127
    Panther, 114, 115, 119, 120
    Parasol, 116, 130, 131
    Perch, 116, 123
    Peto, 116, 125, 126, 127, 140
    Pike, 116, 123, 124
    Pipit, 116, 128, 129
    Pixie, 116, 122, 140
    Plover, 95, 115, 119, 120, 121, 140
    Possum, 121, 122
    Prawn, 116, 129
    Puffin, 115, 119, 120
    Scout (Zepp-Chaser), 115
Percival Gull Four, 135
    Prentice, 11
    Provost (piston), 11, 13, 14
Phoenix engine, 2, 3
Pobjoy engine, 127, 134
Power Jets engines, 59, 60, 137
Pratt & Whitney Hornet engine, 40, 98, 105

Ricardo-Burt engine, 129
Rolls-Royce engines, 4, 12, 33, 39, 41, 45, 46, 48, 52, 54, 57, 58, 60, 62, 77, 81, 87, 91 93 99, 109, 110, 111, 123, 128, 137, 138, 140
Rover W2B/W2B/23 engine, 59, 60, 62

Royal Aircraft Factory SE5a, 15, 16, 26, 29, 31, BE2, 114

Salmson engine, 37
Savoia Marchetti SM79 and SM81, 50
Scottish Aviation Ltd (SAL) Bulldog, 73
Short Belfast, 11
    Bomber, 115
    Kent, 94
    Rangoon, 95
    Seaplane, 115
    Stirling, 11
    Sunderland, 119
Siddeley Puma engine, 33, 81, 83, 84
Sikorsky S-51, 105
    S-58, 108
Slingsby Venture, 11
Sopwith 107, 94
    Camel, 15
    Pup, 79
    Snipe, 27
    1½ Strutter, 15
    Tabloid, 79
Stinson L-5 Sentinel, 15
Sud-Aviation SST, 111
Sunbeam Arab engine, 80, 81
Supermarine Seaplane Racers, 21, 22, 23, 25

Spitfire, 4, 5, 10, 12, 14, 15, 44, 48, 53, 118, 119, 132, 133

V-1 Flying Bomb, 62
Viale engine, 83
Vickers Type 253, 131
    Varsity, 8, 11, 13
    Vespa, 74
    Vildebeest, 29
    Vincent, 44
    Viscount, 4
    Wellesley, 44, 131
    Wellington, 4, 9, 10, 12, 97, 119
Victa Air-Tourer, 10

Webb-Peet Monoplane, 1, 2
    Rotary engine, 2
Westland Wapiti, 34, 37, 88, 95
    Welkin, 57
    Wessex (monoplane), 109
    Whirlwind F.B., 57, 73
    Witch, 35
    Wolseley Viper engine, 82
    Wyvern, 5, 6
Wright engines, vii, 40

Zodiac aeroplane, 71

GENERAL

Advanced Air Striking Force (AASF), 49
Advanced Flying Units (AFUs), 9, 13, 14, 15
Aerial Derby, 19, 20, 82
Aero Show Olympia, 26, 76, 83, 89
Aerospatiale, 111, 112
Air Board, 16
Aircraft and Armament Experimental Establishment (A&AEE), 19, 20, 27, 28, 29, 32, 33, 35, 39, 41, 42, 43, 46, 48, 58, 60, 70, 84, 90, 95, 104, 105, 110, 122, 124, 129
Aircraft Manufacturing Co. (Airco), 16
Aircraft Operating Co., 40, 41
Aircraft Preparation Unit (APU), 73
Aircraft Storage Units (ASUs), 10, 12
Air Experience Flight (AEF), 73
Air Ministry, 18, 20, 26, 27, 29, 31, 32, 34, 35, 36, 38, 42, 43, 45, 46, 47, 48, 53, 54, 60, 61, 62, 73, 84, 87, 89, 92, 95, 99, 119, 121, 123, 124, 128, 129, 130, 131, 132
Air Observer Navigation School/Observer School No. 6, 9
Air Training Corps (ATC) Volunteer Gliding School, 11
Air Transport Auxiliary (ATA), 10
Apsley of Badminton, Lord, 128
Argus, HMS, 18, 119
Armstrong Siddeley, 66, 68, 70, 73
Armstrong Whitworth Aircraft, 64, 70
Army Air Corps, 106
Aston Down, Glos., 10, 15
Atcherley, Flt Lt R.L.R., 28
Australian Flying Corps (AFC), 15
Auty, Geoffrey, 110, 111
Auxiliary Air Force, 33

INDEX

Avery Group, 115, 116
Avions Fairey, 63
A.W. Hawkesley Ltd, 60, 61, 70

Bader, Plt Off Douglas R.S., 29
Barford St John aerodrome, 60
Barnard, F.L., 88
Barnwell, Frank, 78, 79, 81, 82, 87, 89, 92, 95, 96
Bentham, Glos., 52, 60
Bleriot, Louis, vii
Bolas, Harold, 115, 116, 119, 120, 122, 123, 125, 127, 128, 130
Bonham-Carter, Flt Lt D.W., 127
Boscombe Down (*see* A&AEE)
Boulton & Paul, 121
Bowstead, J.G., 5, 6
Brabazon of Tara, Lord, 102, 104
Brazil Straker Co., 93, 94
Bristol Aero Engines Ltd/Bristol Siddeley Engines Ltd, 70, 73
Bristol Aeroplane Co. Ltd, vii, 11, 72, 94, 115
    Armament Division, 76, 89
Bristol Aircraft Ltd, 76
Bristol Tramway & Carriage Co., 71
Bristol University Air Squadron (UAS), 73
Bristol & Wessex Aero Club, 123
British Aerospace (BAe), 73, 76
British Aircraft Corporation (BAC), 76, 110, 111
British Airways, 111, 112
British & Colonial Aeroplane Co. Ltd, 71, 72
British European Airways (BEA), 106, 108
British Messier, 5
British Midland Airways, 10
British Overseas Airways Corporation, 102, 106, 107, 112
British Thomson Houston, 59
British United Air Ferries, 105
Broad, Capt Hubert, 21
Brockworth, Glos., 13, 16, 37, 38, 40, 44, 47, 48, 51, 52, 53, 56, 57, 58, 59, 60, 61, 68, 70
Bruce, Hon Mrs Victor, 134, 135
Burges, Sqn Ldr, 50
Burroughes, Hugh, 16, 36
Busteed, Harry, 79
Butler, Alan, 41

Cambrian Airways, 8, 10
Campbell, Flg Off K., 98
Carbery, Lord, 79
Carter, Larry, 20, 21, 25, 29, 82
Carter, W.G., 57, 59, 62
Centaur, HMS, 109
Central Flying School (CFS), 10, 15, 63
Cheltenham, Glos., 2, 3, 4, 5, 6, 7, 8, 16, 17, 18, 19, 22, 26, 34, 36, 52, 59, 81

Christiaens, Joseph, 76
Cierva, Juan de la, 124, 125
Circuit of Britain Race, 76
Cirencester, Glos., 14, 15
Cirrus Hermes Engineering Co., 133
Coanda, Henry, 77, 78
Cody, S.F., 2
Coliseum Works, Park Row, Bristol, 115
Cornwall Aviation Co., 128
Cosford Aerospace Museum, 110
Cosmos Aero Engine Co., 73, 93, 94
Cotswold Aero Club, 8, 10, 128
Cotswold Gliding Club, 10
Coupe Deutsch Race, 19
Courageous, HMS, 46
Courtney, Frank, 122, 124
Crouch-Bolas Aircraft Co., 116

Dan-Air, 10
Daniels & Co., 16
D'Arcy Grieg, Flg Off D., 24
Daunt, Michael, 9, 59, 62
Davie, Sqn Ldr, 59, 60
Day, Flt Lt M.M., 29
Dean Flg Off, 62
de Haga Haig, Flt Lt 'Rollo', 20, 84
De Havilland Aircraft Co., 39, 54, 69
de Havilland, Geoffrey, 16
Denniston Burney, Lt C., 78
Derby Airways, 10
Dexter, J.R., 5, 6
Dickson, Capt Bertram, 76, 77
Dodge, Francis James, 2
Donaldson, Gp Capt E.M., 63
Doolittle, Lt James H., 22
Dowty, George (later Sir George), 5, 8
Dowty Organization, 5, 6, 7, 43, 47, 48, 66

Eagle, HMS, 108
Elementary & Reserve Flying Training School (E&RFTS), 9, 72

Farman Brothers, 16, 76
Farnborough, 16, 26, 30, 31, 32, 40, 60, 68, 82, 83, 84, 87, 88, 89, 90, 92, 95, 96, 99, 100, 102, 103, 104, 105, 107, 108, 110, 111, 112, 113, 115, 119, 120, 121
Feddon, Roy (later Sir Roy), 73, 75, 85, 94, 95
Fighter Leader School (FLS), 10, 15
Filton, 11, 36, 71, 72, 73, 74, 75, 76, 77, 79, 82, 87, 88, 89, 90, 92, 95, 96, 99, 100, 102, 103, 104, 105, 107, 108, 110, 111, 112, 113, 115, 119, 120, 121
Finnish Air Force, 30, 45, 93
Fleet Air Arm (FAA), 45, 51, 57, 63, 111

INDEX

Fleet Spotter/Reconnaissance Flights, 119, 121
Flight One, 10
Flight Refuelling, 9
Flying Instructors School (FIS), 14, 15
Flying Training Schools (FTSs), 10, 13, 14
Fokker Aircraft Co., 63
Folland, Henry, 16, 17, 18, 19, 20, 21, 23, 26, 27, 29, 31, 33, 34, 35, 40, 41, 42, 46, 47, 48
Folland Aircraft Ltd, 9
Foote, Major Leslie, 82
Frazer-Nash, 117, 119

General Aircraft Ltd, 37
Glider Pilot Regiment, 15
Glorious, HMS, 51
Glosair Ltd, 10
Gloster Aircraft Co. Ltd (GAC), vii, 5, 13, 16, 36, 44, 52, 57, 60, 62, 81
Gloster Trading Estate, 70
Gloucester, 2, 4, 8, 16, 51, 52
Gloucester & Cheltenham School of Flying, 10
Gordon England, E., 77
Grandseigne, Robert, 76
Greenwood, Eric, 4, 5, 63
Grierson, John, 59, 60
Grosvenor Trophy Race, 87, 123

Hafner, Raoul, 105
Hall, Jack, 4
Hall, R., 128
Hancock, Sgt B., 10
Handley Page Transport, 84, 85
Hawker Aircraft Co., 44, 52
Hawker, Harry, 18, 94
Hawker Siddeley Group, 60
Hendy Aircraft Co., 132, 133
Hermes, HMS, 119
Hinkler, Bert, 22, 87
Holt Thomas, George, 16
Hooker, Stanley (later Sir Stanley), 107
Hosegood, C.T.D. 'Socks', 108
Houston, Lady, 21
Hucclecote, Glos., 13, 16, 29, 30, 35, 36, 37, 38, 39, 41, 47, 48, 51, 52, 58, 59, 70
Hunting Aircraft Ltd, 110

Imperial Airways, 84
Imperial War Museum, Duxford, 70
International Aero Show, 40
International Air Tattoo, 11
Intra Airways, 10
Isle of Grain Naval Air Station, 115, 119

James, J.H. 'Jimmie', 18, 19, 20

Kawasaki Aircraft & Engineering Co., 56
Keighley-Peach, Lt C., 126
King's Cup Air Race, 28, 29, 127, 133
Kingsford Smith, Sir Charles, 135
Kinkead, Flt Lt S.M., 23

Larkhill Flying School, 76, 77, 78, 79
Light Aeroplane Competition, 25, 87, 88, 122
Lincoln, Earl of, 44
Longden, David, 36
Lowe, Sqn Ldr, 36
Luftwaffe, 52, 59, 72
Lympne aerodrome, Kent, 25, 26, 87, 122

Mackenzie-Richards, Flg Off, 28
MacMillan, Capt Norman, 121
Malcolm, Wg Cdr H.G., 97
Malta G.C., 9, 50, 51, 56, 96, 98, 99, 100
Marsh, H.A., 105
Marine Aircraft Experimental Establishment (MAEE), 21, 22, 124, 126, 130
Martin Baker Ltd, 63
Martin, Wg Cdr R.F. 'Dicky', 67, 68
Martlesham (*see* A&AEE)
Martyn, A.W., 36
Martyn, H.H., 3, 16
McKenna, Frank, 55
Miles, F.G. 'Freddie', 134
Ministry of Aircraft Production, 60
Ministry of Aviation, 10, 111
Ministry of Defence (MoD), 11
Ministry of Supply (MoS), 73, 104, 105, 106, 108
Mitchell, Flt Lt A.B., 44
Mitchell, R.J., 21
Mitsubishi Aircraft Co., 32
Monospar Wing Co., 37
Moreton Valence, Glos., 13, 62, 63, 66, 70

Nakajima Aircraft Co., 32, 33, 92
Napier, Carril S., 133
Nash & Thompson Ltd, 119, 132
New Zealand Permanent Air Force (NZPAF), 29
Nieuport & General, 18
Noakes, Flt Lt J., 129
Norwegian Campaign, 51

Orchard, Flg Off A.T., 127
Overseas Aircraft Delivery Squadron, 12
Overseas Aircraft Preparation Unit (OAPU), 73

Palmer Brakes, 5, 6
Palmer Brothers, 2, 3, 4
Paris Air Show, 73, 77
Parnall, George & Co., vii, 13, 115, 116, 117, 119, 120, 121, 122, 123, 124, 125, 127, 133, 135

INDEX

Parnall Aircraft Ltd, 13, 117, 119, 132
Parnall and Sons Ltd, vii, 115, 119
Peck, Guy, 16
Pegg, Flt Lt A.J. 'Bill', 95, 102, 107
Percival, Capt E.W., 133, 135
Phoenix Radial Rotary Motor Co., 2
Pierson, Rex, 75
Pobjoy, D.R., 127
Pope, Flt Lt H.N. 'Poppy', 129
Preston, Sir Walter, 28
Prier, Pierre, 76

RAE, Bedford, 110, 111
Railway Air Services, 9
Reid, W.T., 82
Reserve Flying School (RFS), 72, 73, 87
Rolls-Royce, 70
Ross, Flt Lt, 68
Rotol Airscrews, 4, 7, 9, 10, 54, 57, 102, 103
Royal Aero Club, 20, 25, 122
Royal Aircraft Establishment (RAE, see Farnborough)
Royal Aircraft Factory (see Farnborough)
Royal Air Force (RAF), 81, 92, 93, 96, 100, 108, 113, 116, 126, 132, 133
   Air Sea Rescue Service, 106
   Calshot, 23, 24, 25, 35
   Coastal Command, 106
   Colerne, 99
   Cranwell, 11, 13, 21, 59
   Delhi Communication Flight, 52
   Fairford, 11, 111
   Fighter Command, 10, 63, 72, 113
   44 Group 9, 10
   Hendon, 19, 25, 26, 27, 35, 41, 42, 48, 121, 122
   High Speed Flight, 20, 23, 25
   Kemble, 11, 69
   Leuchars, 70
   Little Rissington, 10, 11
   Maintenance Units (MUs), 10, 11, 12, 13, 69
   Manston, 62
   Museum, 51
   Odiham, 109
   Operational Training Units (OTUs), 10, 14, 15
   Police Dog Training School, 10
   Police Wing No. 1, 10
   Red Arrows Aerobatic Team, 13
   Service Flying Training Schools (SFTs), 15
   South Cerney, 13
   Staverton (see Staverton)
   2nd Tactical Air Force, 13, 57, 62
   Tangmere, 72
   Transport Command, 12
Royal Australian Air Force (RAAF), 45, 63, 96, 98
Royal Australian Navy (RAN), 106

Royal Canadian Air Force (RCAF), 98
Royal Flying Corps (RFC), 72, 77, 78, 80, 81, 82, 116
Royal Naval Air Service (RNAS), 2, 77, 78, 80, 82
Royal Navy, 50, 126
Royal New Zealand Air Force (RNZAF), 29

Saint, Capt Howard, 30, 37, 40, 41, 44
Savages Ltd, 16
Sayer, Fl Lt P.E.G. 'Gerry', 7, 44, 47, 48, 58, 59
Scarf, Sqn Ldr A.S.K., 98
Schneider Trophy, 20, 21, 22, 23
Scott Brothers, 1
Shaw, C.R.L., 88
Short Brothers, 95, 107
Shuttleworth Trust, 51, 128
Silver City Airways, 105
Skyfame Museum, 10
Smart, Air Marshal, 133
Smiths Industries, 8, 10
Society of British Aircraft Constructors (SBAC) 12, 106, 108
South African Air Force, 41, 45
Stainforth, Flt Lt G.H., 24, 25, 28
Staverton airfield, Glos., 4, 8, 9, 10, 106
Steel Wing Co., 36, 82
Steiger, H.J., 37
Stocken, Rex, 35, 37
Sud-Aviation, 111
Sutherland, Duke of, 25, 122
Swedish Army Air Service, 27
   Air Force Museum, 51

Tactical Exercise Unit (TEU), 10, 15
Tomkins, J.W., 30
Trubshaw, Brian, 111
Twiss, Peter, 110

United States Air Force (USAF), 11, 13
United States Army Air Force (USAAF), 15
Uwins, C.F., 75, 84, 87, 88, 89, 95, 104

Waters, Capt S.J., 34
Waterton, Sqn Ldr W.A. 'Bill', 60, 66, 67
Webb-Peet & Co., 1
Webster, Flt Lt S.N., 23
Westland Aircraft Ltd, 109, 121, 132
White, Sir George, Bt., 71
Whittle, Frank (later Sir Frank), 58, 59
Wilson, Gp Capt H.J., 62
World Air Speed Record, 19, 20, 23, 25

Yate, Glos., 115, 116, 117, 118, 119, 121, 122, 124, 125, 126, 127, 128, 129, 130, 131, 132, 133, 134, 135